NURBS

NURBS
From Projective Geometry to Practical Use
Second Edition

Gerald E. Farin
Arizona State University
Tempe, Arizona

CRC Press
Taylor & Francis Group
Boca Raton London New York

CRC Press is an imprint of the
Taylor & Francis Group, an **informa** business

AN A K PETERS BOOK

First published 1999 by A K Peters, Ltd.

Published 2018 by CRC Press
Taylor & Francis Group
6000 Broken Sound Parkway NW, Suite 300
Boca Raton, FL 33487-2742

ISBN 13: 978-1-56881-084-3 (hbk)

**Visit the Taylor & Francis Web site at
http://www.taylorandfrancis.com**

**and the CRC Press Web site at
http://www.crcpress.com**

Cover illustration from *The Complete Woodcuts of Albrecht Dürer*, edited
by Dr. Willi Durth, Dover Publications, Inc., New York, 1963. Plate 338
taken from *Illustrations to the Treatise on Measurement,* Unterweysung
der Messung, Nürnberg, 1525. A man drawing a lute.

Library of Congress Cataloging-in-Publication Data

Farin, Gerald E.
    NURBS : from projective geometry to practical use / Gerald E.
    Farin. -- 2nd ed.
      p.   cm.
    Rev. ed. of: NURB curves and surfaces. c1995.
    Includes bibliographical references (p.   -) and index.
    ISBN 1-56881-084-9
    1. Spline theory--Data processing. 2. Computer-aided design.
  I. Farin, Gerald E. NURB curves and surfaces. II. Title.
QA224.F37 1999
511'.42--dc21   99-12045
                                                    CIP

For Dianne

Contents

Preface

Computers have become an integral part of today's design and manufacturing technology. Consequently, the field now employs sophisticated mathematical methods: mathematics is the only language "understood" by computers. One aspect of the design process — modeling complex parts such as car bodies or airplane fuselages — has given rise to techniques known as Bézier curves and Coons methods. The corresponding theories, together with that of B-splines, constitutes what is now referred to as "curve and surface design."

Until quite recently, different companies used different variations of these basic techniques: General Motors used Gordon surfaces, Ford used Coons' patches, and Chrysler used Chebychev polynomials, for example. This diversity of methods led to communication problems: a product definition in one form is typically not translatable into another system. Thus, a *geometry standard* was called for, and it was found in the form of NURBS (Non-Uniform Rational B-Splines, a term that will be explained later).

I find it very instructive to look at development at the Boeing Co.: during the 1950s, engineers "computerized" existing design and drafting methods that were all based on the use of conics.[1] In the late 1950s,

---

[1] For an excellent exposition of these methods, see the book by Liming [102]. This book is the first to introduce algorithms into the design process — although in those days, CAD would probably have been translated as "Calculator Aided Design."

J. Ferguson developed a spline package for Boeing's design software. Both packages were independently used, and eventually it was discovered that they were essentially incompatible. That is to say, a spline is not a conic, and a conic is not a spline, with the somewhat trivial exception of the parabola. Management realized that these different methods had to be unified, and thus NURBS were born: they encompass both splines and conics.

The first appearance of NURBS is in K. Vesprille's PhD thesis [151], where he makes heavy use of homogeneous coordinates, an approach that goes back to S. Coons and R. Forrest. Riesenfeld [130] realized early that using homogenous coordinates mandates working with projective geometry. This book follows these ideas and attempts to base the theory of NURBS firmly in projective geometry. This way, we gain more insight into theoretical issues as well as practical ones.

After a general outline of projective geometry, we introduce conics through a classical projective definition. Using the concept of cross ratios, we arrive at the de Casteljau algorithm for conics. Then we cover areas such as rational Bézier curves curves, NURBS curves and surfaces, triangular patches, Gregory patches, and more. The book closes with some practical examples, including a discussion of the IGES NURBS data specifications.

I have taught the material in this book in one-semester classes, typically at the beginning graduate level. The prerequisites are linear algebra, calculus, and some basic computer graphics. The latter is needed for sucessful completion of programming assignments, which I find indispensable for a true comprehension of the subject. Such exercises, both practical and theoretical, are included at the end of each chapter.

For this second edition, several changes have been incorporated. The most visible ones are the gray-shaded circle figures such as Fig. 7.14. They are intended to highlight some of the phenomena that arise when dealing with objects in projective geometry. These figures were all generated by PostScript programs, which are themselves generated by a few C routines. All of the figures, as well as future errata and additions are available on the web site `http://eros.eas.asu.edu/~farin/rbook/rbook_home.html`

Another group of figures, such as Fig. 11.15, were generated using GEOMVIEW [2]. These simply give more visual intuition for working with NURBS. The corresponding files are also available on the above web site.

Essentially all of the chapters have been expanded. One new chapter has been added, namely Chapter 10; it appears that Pythagorean Hodograph curves and their relatives are playing an increasingly important role.

---

[2]To download GEOMVIEW, visit `http://www.geom.umn.edu/software/geomview/`

This book would not have been possible without the help of many friends and colleagues. I would like to thank R. Barnhill, K. Choodamani, L. Gross, D. Jung, P. Kashyap, T. Kim, A. Razdan, A. Palacios, T. Sederberg, R. Smith, A. Swimmer, H. Wolters, and A. Worsey for their many contributions. The most support, however, came from D. Hansford, who carefully monitored the progress of this work. I also acknowledge support from the National Science Foundation through Grant No. DM9123527 and a grant of the German Research Council (DFG).

Gerald Farin
Tempe, Arizona

1

The Projective Plane

1.1 Motivation

Most of us have studied geometry to some extent once, and in most cases, it was *euclidean geometry*. That geometry is perfectly adequate to handle objects such as points and lines, circles and planes, and much more. In fact, it is used successfully in the computer aided construction of cars, ships, and airplanes — the process known as CAD, or Computer Aided Design.

As we shall see later, *projective geometry* is the natural setting for NURBS: many of their intrinsic properties are much easier understood in a projective context.

Euclidean geometry is an indispensable tool for some applications, but it has its shortcomings in other areas. If we are to deal with how we *see* things, the euclidean viewpoint becomes less advantageous. Looking up at high rise buildings, we perceive parallel lines as converging, and euclidean geometry can't cope with that. In projective geometry, there are no parallel lines — it is therefore the geometry of choice for those working with perceived images. Architects use projective geometry when drawing a building as it would appear to an observer; computer graphics programmers use it when they model "realistic" scenes.

The roots of projective geometry go back to the late middle ages; and they were not in mathematics, but in the arts. Medieval painters neglected

Figure 1.1. Perception of an object: the observer "sees" the object's projection into a plane. Woodcut by A. Dürer, 1525.

the concept of realistic perspective for religious reasons. It was the advent of the renaissance that led artists to study the techniques necessary for realistic rendering. Among them, we find names such as P. Uccello[1], Leonardo da Vinci[2], and Albrecht Dürer[3]. First mathematical excursions into the field of perspective were carried out by B. Pascal[4] and G. Desargues[5] in the seventeenth century. Since then, perspective was an indispensable tool for artists, architects, and engineers; but it took until the nineteenth century that it became the object of formal mathematical inquiry, culminating in the theory of projective geometry. Some of the names associated with this breakthrough are J. Poncelet, K. von Staudt, J. Steiner and F. Moebius.

Before giving formal definitions in the next section, let us motivate the underlying principle of projective geometry. Remember, this geometry was invented to model how we perceive objects. So let's assume that an

---

[1] The cover of the journal "Computer Aided Geometric Design" is one of Uccello's studies of perspective renditions.

[2] Da Vinci, L.

[3] Dřer, A.

[4] Pascal, B.

[5] Desargues, G.

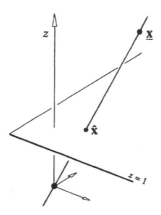

Figure 1.2. Perception of a point: all points on the shown line appear identical to an observer at the origin.

observer is looking at a three-dimensional object. The observer's perceived image of the object is two-dimensional: it is the projection of the object into a plane, as shown in Fig. 1.1.

This figure shows the scientific/engineering approach that Dürer took in order to master perspective: he used wires to record the perceived position of points that were marked on an object. Just replace "wires" by "coordinates" and "marked" by "digitized," and you see how close this approach was to our modern way of thinking!

Now for a more formal treatment: place the observer at the origin of a 3D orthogonal coordinate system, and let $z = 1$ denote the plane into which we project. Let us make our idealized observer look at some 3D point $\mathbf{x}$. The perceived image is $\mathbf{x}$'s projection into the plane $z = 1$. But many more 3D points project onto the same image: all these points, located on the straight line through the origin and $\mathbf{x}$, are indistinguishable to our observer! Fig. 1.2 illustrates this example. In projective geometry, since we can't tell those points apart anyway, they are treated as *one point*.

This idea — only treat as an entity what you can identify after you project it into the plane — is carried over to all geometrical objects. Let's take the case of straight lines, or lines for short. Referring to Fig. 1.3, we see that all lines contained in the indicated plane project onto the same line in the plane $z = 1$. Consequently, they are treated as being identical! The plane containing the 3D lines is characterized by its normal vector, and so this normal vector is used to denote the equivalent lines.

Our treatment of projective geometry will not be exhaustive — we just need it as a tool for the description of rational curves and surfaces. For more

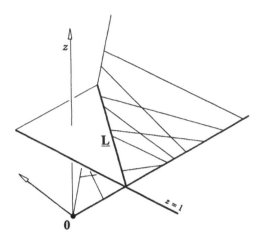

Figure 1.3. Perception of a line: all lines in the indicated plane appear identical to an observer at the origin.

details, the interested reader should consult the texts by Boehm/Prautzsch [26], Coxeter [37], Hilbert/Cohn-Vossen [86], Pedoe [115], Penna/Patterson [116], or Struik [144].

1.2   The Projective Plane

The intuitive concepts from the preceding section lead to the following formal definition of the projective plane $I\!P^2$:

These definitions invite several comments:

1. The coordinate column $[0,0,0]^{\mathrm{T}}$ designates a point, but not a *location*: as it lies on every line, its location is indeterminate. Similarly, the coordinate row $[0,0,0]$ designates a line, but its location is indeterminate as every point lies on it.

2. We had used the plane $z = 1$ to introduce $I\!P^2$, but now points with $z = 1$ are not special. We will treat them as special later when we talk about affine geometry.

3. The projective plane does not possess "natural" coordinate axes; in particular, it does not possess an origin. (In particular, $[0,0,0]^{\mathrm{T}}$ is not a candidate for an origin, since it does not even designate a location.)

4. The sign "=" denotes algebraic equivalence, while "≐" denotes geometric equivalence.

5. We may change point *coordinates* to any nonzero multiple without changing the point's *location*: $\underline{x} \doteq \underline{y}$ implies $\underline{x} = \alpha\underline{y}$. But we may *not* change point coordinates within expressions: $\underline{x} = \underline{y} + \underline{z}$ does not imply $\underline{x} = \underline{y} + \alpha\underline{z}$. This is also true for line coordinates.

6. Geometrically, $I\!P^2$ is a two-dimensional object. Using three coordinates to represent its points (or lines) will lead to a very elegant mathematical treatment.

Let us now turn to Fig. 1.4 for an illustration of (1.1). The line $\underline{L}$ is represented by the affine vector l, which is the normal vector of the plane containing $\underline{L}$ and the origin. This plane also contains the affine vector **p** from the origin to $\underline{p}$. This vector is thus orthogonal to the normal vector, as stated in (1.1).

The **projective plane** $I\!P^2$ consists of points $\underline{p}$, denoted by coordinate columns

$$\underline{p} = \begin{bmatrix} p_1 \\ p_2 \\ p_3 \end{bmatrix}$$

and lines $\underline{L}$, denoted by coordinate rows

$$\underline{L} = [l_1, l_2, l_3].$$

All nonzero multiples $\alpha\underline{p}$ denote the same location, and we define $\underline{p} \doteq \alpha\underline{p}$. Similarly, all nonzero multiples $\alpha\underline{L}$ are treated as the same line, and we define $\underline{L} \doteq \alpha\underline{L}$.
A point $\underline{p}$ is on a line $\underline{L}$ if

$$\underline{L}\underline{p} = 0, \tag{1.1}$$

i.e., if the dot product of their coordinates vanishes.

1.3   Points and Lines

Intuitively, two distinct points define a line through them. If our definition of the projective plane is meaningful, then it should support this notion.

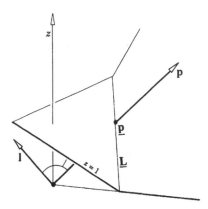

Figure 1.4. A point on a line: If $\underline{p}$ is on $\underline{L}$, then p must be orthogonal to l.

It does indeed do that: let $\underline{a}$ and $\underline{b}$ be two distinct points in $I\!\!P^2$. Can we find a line that passes through them? If that line is denoted by $\underline{L}$, then $\underline{a}$ lies on it:

$$\underline{L}\underline{a} = 0,$$

and so does $\underline{b}$:

$$\underline{L}\underline{b} = 0.$$

It now takes nothing but elementary linear algebra to conclude that

$$\underline{L} = \underline{a} \wedge \underline{b}, \tag{1.2}$$

with " $\wedge$ " denoting the cross or vector product. It is computed using the following determinants:

$$\underline{L} = \left[\begin{array}{ccc} \left| \begin{array}{cc} a_2 & b_2 \\ a_3 & b_3 \end{array} \right|, & - \left| \begin{array}{cc} a_1 & b_1 \\ a_3 & b_3 \end{array} \right|, & \left| \begin{array}{cc} a_1 & b_1 \\ a_2 & b_2 \end{array} \right| \end{array} \right]. \tag{1.3}$$

If we replace $\underline{a}$ by a nonzero multiple $\alpha\underline{a}$, we have not changed $\underline{a}$'s location. Consequently, $\alpha\underline{a}$ and $\underline{b}$ should define the same line as did $\underline{a}$ and $\underline{b}$. And they do: $\alpha\underline{a} \wedge \underline{b} = \alpha(\underline{a} \wedge \underline{b}) = \alpha\underline{L} \hat{=} \underline{L}$.

Now for the corresponding concept for lines: the intersection of two lines should define a point. So let $\underline{L}$ and $\underline{M}$ denote two lines. If they intersect in a point $\underline{x}$, then $\underline{x}$ must satisfy

$$\underline{L}\underline{x} = 0$$

and also

$$\underline{\mathbf{M}}\underline{\mathbf{x}} = 0.$$

Hence $\underline{\mathbf{x}}$ can be found as a cross product:

$$\underline{\mathbf{x}} = \underline{\mathbf{L}} \wedge \underline{\mathbf{M}}, \tag{1.4}$$

which is quite similar to (1.2). For some examples, see Fig. 1.7.

Any two distinct points define a line — that seems fairly obvious. But if any two distinct lines intersect in a point, are we not excluding parallel lines? We are indeed: there are no parallel lines in projective geometry!

1.4   Projective Line Coordinates

Two points define a line — but surely there must be more than two points on a line. If $\underline{\mathbf{a}}$ and $\underline{\mathbf{b}}$ define a line, then what does it take for a third point $\underline{\mathbf{x}}$ to be on that line? Since the line is given by $\underline{\mathbf{a}} \wedge \underline{\mathbf{b}}$, the point $\underline{\mathbf{x}}$ must satisfy

$$[\underline{\mathbf{a}} \wedge \underline{\mathbf{b}}]\underline{\mathbf{x}} = 0,$$

which is equivalent to

$$\det[\underline{\mathbf{a}}, \underline{\mathbf{x}}, \underline{\mathbf{b}}] = 0. \tag{1.5}$$

Equation (1.5) gives a necessary and sufficient condition for three distinct points to lie on the same line, or to be *collinear*.

There is a similar condition for three lines going through one point. Three such lines are called *concurrent*. Using the same approach as for (1.5), we obtain:

Three distinct lines $\underline{\mathbf{L}}, \underline{\mathbf{X}}, \underline{\mathbf{M}}$ are concurrent if and only if

$$\det[\underline{\mathbf{L}}, \underline{\mathbf{X}}, \underline{\mathbf{M}}] = 0. \tag{1.6}$$

If three points are collinear, then their coordinate columns are linearly dependent, as expressed by (1.5). We may thus write one point as a linear combination of the other two:

$$\underline{\mathbf{x}} = \alpha\underline{\mathbf{a}} + \beta\underline{\mathbf{b}}, \tag{1.7}$$

where α and β are suitable real numbers. By letting α and β trace out all reals, $\underline{\mathbf{x}}$ will assume all point locations on the line $\underline{\mathbf{L}}$ through $\underline{\mathbf{a}}$ and $\underline{\mathbf{b}}$. Thus each point on the line is determined by two numbers α and β or any multiple thereof. The pair $[\alpha, \beta]$ is referred to as *projective coordinates* of the point $\underline{\mathbf{x}}$ with respect to the line defined by $\underline{\mathbf{a}}$ and $\underline{\mathbf{b}}$.

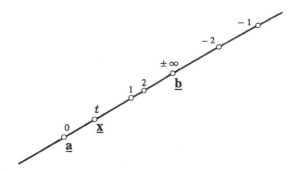

Figure 1.5. Tracing out a line: the point $\underline{\mathbf{x}}$ depends on t, and its location is shown for various values of t.

We can also make $\underline{\mathbf{x}}$ trace out $\underline{\mathbf{L}}$ by letting it depend on one parameter only. Let us set

$$\underline{\mathbf{x}} = \underline{\mathbf{a}} + t\underline{\mathbf{b}}, \tag{1.8}$$

again with t being any real number. Fig. 1.5 illustrates *how* $\underline{\mathbf{x}}$ traces out the line $\underline{\mathbf{L}}$ as a function of the parameter t. Note that the point $\underline{\mathbf{b}}$ corresponds to the two parameter values $t = \pm\infty$.

When we refer to a line as the collection of all points on it, we call it a *range*. Equation (1.8) gives an example of a *parametrization* of a range. By picking different points $\underline{\mathbf{a}}$ and $\underline{\mathbf{b}}$ on $\underline{\mathbf{L}}$, we produce different parametrizations. Even changing the coordinates of $\underline{\mathbf{b}}$ to $c\underline{\mathbf{b}}$ will result in a different parametrization of $\underline{\mathbf{L}}$! See Section 1.12 for more on this topic.

The same principles hold true for lines: if $\underline{\mathbf{L}}, \underline{\mathbf{X}}$, and $\underline{\mathbf{M}}$ are concurrent, i.e., all containing a point $\underline{\mathbf{p}}$, then we can write one of them as a linear combination of the others:

$$\underline{\mathbf{X}} = \alpha\underline{\mathbf{L}} + \beta\underline{\mathbf{M}}. \tag{1.9}$$

This will look strange to those who encounter projective geometry for the first time — but clearly a linear combination of two coordinate rows yields another row, and thus denotes a line. Again, we refer to $[\alpha, \beta]$ as *projective coordinates* in the *pencil* of lines through $\underline{\mathbf{p}}$.

As α and β assume all possible real values, (1.9) produces all lines that pass through the intersection of $\underline{\mathbf{L}}$ and $\underline{\mathbf{M}}$. Again, we may wish to express $\underline{\mathbf{X}}$ as depending on one parameter only; then we write

$$\underline{\mathbf{X}} = \underline{\mathbf{L}} + t\underline{\mathbf{M}}. \tag{1.10}$$

Equation (1.10) gives an example of a *parametrization* of a pencil. Fig. 1.6 illustrates.

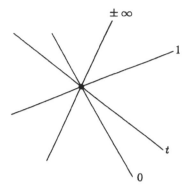

Figure 1.6. Tracing out a pencil: the line $\underline{X}$ depends on t, and its location is shown for various values of t.

The following "trick" will be helpful in the future. Let $\underline{a}, \underline{x}, \underline{b}$ be three distinct collinear points, such that (1.7) holds. We may scale the components of each point (thus not changing locations!) in a suitable way, so that we can write

$$\underline{x} = \underline{a} + \underline{b} \qquad (1.11)$$

with suitably redefined $\underline{a}$ and $\underline{b}$. Thus if we are given any three distinct collinear points, they may always be rewritten such that they are in relation (1.11). Example 1.1 gives a numerical demonstration.

In terms of projective coordinates on the line, the three points $\underline{a}, \underline{x}, \underline{b}$ may be denoted by $[1, 0], [1, 1], [0, 1]$, respectively. This works only once: if we have assigned these projective coordinates to three points on a line, then the coordinates of all other points on the line are fixed! That is why we say that three distinct points on a line constitute a *projective reference frame* for a line.

Let three collinear points be given by

$$\underline{a} = \begin{bmatrix} 1 \\ 0 \\ 1 \end{bmatrix}, \underline{x} = \begin{bmatrix} 8 \\ 0 \\ 5 \end{bmatrix}, \underline{b} = \begin{bmatrix} 2 \\ 0 \\ 1 \end{bmatrix},$$

such that $\underline{x} = 2\underline{a} + 3\underline{b}$. Now redefine $\underline{a} \leftarrow 2\underline{a}$ to and $\underline{b} \leftarrow 3\underline{b}$. With these new definitions, we have $\underline{x} = \underline{a} + \underline{b}$.

Example 1.1. Adjusting point components.

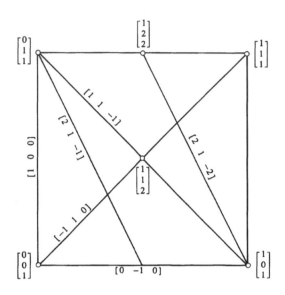

Figure 1.7. Projective coordinates: some examples of points and lines.

Let $\underline{\mathbf{a}}_1, \underline{\mathbf{a}}_2, \underline{\mathbf{a}}_3$ and $\underline{\mathbf{b}}_1, \underline{\mathbf{b}}_2, \underline{\mathbf{b}}_3$ be two projective reference frames for some line $\underline{\mathbf{L}}$. We may always set $\underline{\mathbf{a}}_2 = \underline{\mathbf{a}}_1 + \underline{\mathbf{a}}_3$ by properly rescaling all coordinates on the line. In general, we will then not have that $\underline{\mathbf{b}}_2 = \underline{\mathbf{b}}_1 + \underline{\mathbf{b}}_3$. But if this *is* the case, we call the two frames *equivalent*. In particular, both reference frames will have the same point corresponding to $t = \infty$.

The reader is invited to verify that the corresponding statements for lines and pencils also hold. Several points and lines, together with their coordinates, are shown in Fig. 1.7.

1.5 Projective Plane Coordinates

Another interesting configuration is given by considering three noncollinear points $\underline{\mathbf{a}}, \underline{\mathbf{b}}, \underline{\mathbf{c}}$ forming a triangle and a fourth point $\underline{\mathbf{d}}$. We can always write $\underline{\mathbf{d}}$ as

$$\underline{\mathbf{d}} = \underline{\mathbf{a}} + \underline{\mathbf{b}} + \underline{\mathbf{c}} \tag{1.12}$$

by suitably redefining $\underline{\mathbf{a}} \leftarrow \alpha\underline{\mathbf{a}}$, $\underline{\mathbf{b}} \leftarrow \beta\underline{\mathbf{b}}$, and $\underline{\mathbf{c}} \leftarrow \gamma\underline{\mathbf{c}}$. All we need to do is solve the linear system

$$\underline{\mathbf{d}} = \alpha\underline{\mathbf{a}} + \beta\underline{\mathbf{b}} + \gamma\underline{\mathbf{c}}$$

for α, β, γ. This system has a unique solution since $\underline{\mathbf{a}}, \underline{\mathbf{b}}, \underline{\mathbf{c}}$ were assumed to be non-collinear. Further rescaling and redefinition of point coordinates

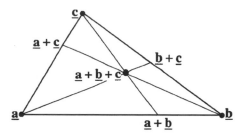

Figure 1.8. Triangle configurations: this labeling of four arbitrary points can always be achieved.

will always allow us to assume that our points are in the relation depicted in Fig. 1.8.

Note that $\underline{d}$ can be anywhere, as long as it is not on one of the triangle edges. In particular, $\underline{d}$ does not have to be "inside" the triangle.[6]

Once four points, no three of them collinear, have been fixed, and coordinates have been assigned to them, the coordinates of every point in the projective plane are determined: we say that the four points constitute a *projective reference frame* for $I\!\!P^2$. See also Problems. Fig. 1.9 illustrates the structure of the corresponding coordinate system.

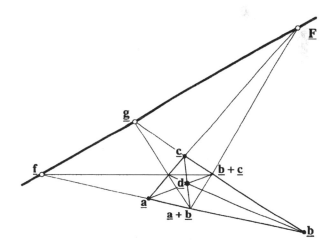

Figure 1.9. Projective coordinate systems: four points form a projective reference frame and determine a fundamental line.

---

[6]See Section 1.11 for more details on this topological aspect of the projective plane.

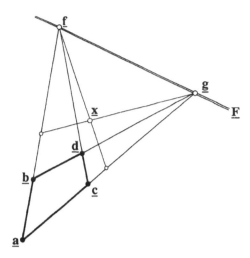

Figure 1.10. Projective coordinate systems: four points may be used to generate a quadrilateral coordinate system.

Without loss of generality, we may always assume

$$\underline{a} = \begin{bmatrix} 1 \\ 0 \\ 0 \end{bmatrix}, \underline{b} = \begin{bmatrix} 0 \\ 1 \\ 0 \end{bmatrix}, \underline{c} = \begin{bmatrix} 0 \\ 0 \\ 1 \end{bmatrix}, \underline{d} = \begin{bmatrix} 1 \\ 1 \\ 1 \end{bmatrix}$$

for any four points $\underline{a}, \underline{b}, \underline{c}, \underline{s}$.

There are three centers of pencils corresponding to the differences of $\underline{a}, \underline{b}$, and $\underline{c}$; we denote them by $\underline{f} = [1, -1, 0]^T, \underline{g} = [0, -1, 1]^T$, and $\underline{h} = [-1, 0, 1]^T$. It is easy to see that they are collinear; we call the line that contains them the *fundamental line* of our coordinate system. It will be used in the treatment of quadrics in Section 13.3.

If we draw the fourth point $\underline{d}$ as being "outside" the triangle spanned by $\underline{a}, \underline{b}, \underline{c}$, then we obtain a network as shown in Fig. 1.10. This network is called a *Moebius net*, see [26]. We also make the stipulation that $\underline{b} = \underline{a} + \underline{f}$ and that $\underline{c} = \underline{a} + \underline{g}$, where $\underline{f}$ and $\underline{g}$ are defined[7] as shown in Fig. 1.10. We call the line $\underline{F} = \underline{f} \wedge \underline{g}$ the *fundamental line* of our Moebius net.

A Moebius net assigns two coordinates u, v to every point $\underline{x}$:

$$\underline{x} = \left[[\underline{a} + u\underline{g}] \wedge \underline{f} \ big \right] \wedge [\underline{a} + v\underline{f}] \wedge \underline{g}. \tag{1.13}$$

In particular, $\underline{b}$ corresponds to $(0, 1)$ whereas $\underline{d}$ corresponds to $(1, 1)$, and $\underline{a}$ corresponds to $(0, 0)$.

---

[7]This definition is different from that of Fig. 1.9!

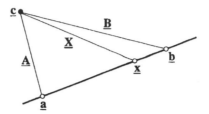

Figure 1.11. Ranges and pencils: collinear points satisfy the same relation as concurrent lines through them.

1.6 Ranges and Pencils

There is an interesting interplay between range and pencil configurations: let $\underline{a}, \underline{x}, \underline{b}$ be three collinear points such that

$$\underline{x} = u\underline{a} + v\underline{b}.$$

Let $\underline{c}$ be a point not on the line through $\underline{a}$ and $\underline{b}$ and define three lines $\underline{A}, \underline{X}, \underline{B}$ by setting $\underline{A} = \underline{a} \wedge \underline{c}$, etc. Then the lines satisfy the same relationship as the points do, i.e.,

$$\underline{X} = u\underline{A} + v\underline{B}. \tag{1.14}$$

For a proof, we refer to Fig. 1.11 and write out $\underline{X}$'s definition:

$$
\begin{aligned}
\underline{X} &= [u\underline{a} + v\underline{b}] \wedge \underline{c} \\
&= u(\underline{a} \wedge \underline{c}) + v(\underline{b} \wedge \underline{c}) \\
&= u\underline{A} + v\underline{B},
\end{aligned}
$$

thus asserting (1.14).

1.7 Pappus' Theorem

This is one of the most fundamental theorems on the projective plane, and will be used in many constructions later on. It can be stated as follows:

Let $\underline{L}$ and $\underline{M}$ be two distinct lines, let $\underline{a}_1, \underline{a}_2, \underline{a}_3$ be three distinct points on $\underline{L}$, and let $\underline{b}_1, \underline{b}_2, \underline{b}_3$ be three distinct points on $\underline{M}$. Then the three points

$$
\begin{aligned}
\underline{P}_1 &= [\underline{a}_2 \wedge \underline{b}_3] \wedge [\underline{a}_3 \wedge \underline{b}_2], \\
\underline{P}_2 &= [\underline{a}_1 \wedge \underline{b}_3] \wedge [\underline{a}_3 \wedge \underline{b}_1], \\
\underline{P}_3 &= [\underline{a}_1 \wedge \underline{b}_2] \wedge [\underline{a}_2 \wedge \underline{b}_1]
\end{aligned}
\tag{1.15}
$$

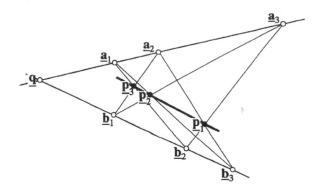

Figure 1.12. Pappus' theorem: the indicated (solid) points are collinear.

are collinear, as illustrated in Fig. 1.12. The line that they all lie on is called the *Pappus line*.

For a proof of this theorem, assume that $\underline{q}$ is the intersection of the two lines $\underline{L}$ and $\underline{M}$. If necessary, we can rescale the coordinates of all involved points such that we have four points on $\underline{L}$ as follows:

$$\underline{q}, \quad \underline{a}_1, \quad \underline{a}_2 \hat{=} \underline{q} + \underline{a}_1, \quad h a_3 \hat{=} \underline{q} + \lambda \underline{a}_1$$

and four points on $\underline{M}$:

$$\underline{q}, \quad \underline{b}_1 \hat{=} \underline{q} + \mu \underline{b}_3, \quad \underline{b}_2 \hat{=} \underline{q} + \underline{b}_3, \quad \underline{b}_3.$$

Example 1.1 shows how to achieve this.

Let us now obtain analytic expressions for the points $\underline{p}_i$. Since $\underline{p}_1$ is the intersection of the two lines $\underline{a}_2 \wedge \underline{b}_3$ and $\underline{a}_3 \wedge \underline{b}_2$, we can write

$$\underline{p}_1 = r(\underline{q} + \underline{a}_1) + s\underline{b}_3 = t(\underline{q} + \underline{b}_3) + u(\underline{q} + \lambda \underline{a}_1).$$

Since the points $\underline{q}, \underline{a}_1, \underline{b}_3$ are assumed to be non-collinear, we can compare coefficients:

$$r = t + u,$$
$$r = u\lambda,$$
$$s = t.$$

A common factor in the r, s, t, u does not matter, so we set $u = 1$ and obtain $r = \lambda$ and $s = t = \lambda - 1$. Thus

$$\underline{\mathbf{P}}_1 = \lambda \underline{\mathbf{q}} + \lambda \underline{\mathbf{a}}_1 + (\lambda - 1)\underline{\mathbf{b}}_3.$$

Using the same technique, we also find

$$\underline{\mathbf{P}}_2 = \lambda \underline{\mathbf{a}}_1 - \mu \underline{\mathbf{b}}_3$$

and

$$\underline{\mathbf{P}}_3 = \mu \underline{\mathbf{q}} + (\mu - 1)\underline{\mathbf{a}}_1 + \mu \underline{\mathbf{b}}_3.$$

In order to show that $\underline{\mathbf{P}}_1, \underline{\mathbf{P}}_2, \underline{\mathbf{P}}_3$ are indeed collinear, we combine their representations:

$$\begin{bmatrix} \underline{\mathbf{P}}_1 \\ \underline{\mathbf{P}}_2 \\ \underline{\mathbf{P}}_3 \end{bmatrix} = \begin{bmatrix} \lambda & \lambda & \lambda - 1 \\ 0 & \lambda & -\mu \\ \mu & \mu - 1 & \mu \end{bmatrix} \begin{bmatrix} \underline{\mathbf{q}} \\ \underline{\mathbf{a}}_1 \\ \underline{\mathbf{b}}_3 \end{bmatrix}.$$

The determinant of the 3×3 matrix is zero, and so Pappus' theorem is proved.

1.8   Menelaos' Theorem

Pappus' theorem will be important when we construct conics later. Another theorem, *Menelaos' theorem*, will be used to prove symmetry properties of many constructive algorithms, and its importance was already realized by de Casteljau [43].

Referring to Figure 1.13, let

$$\underline{\mathbf{a}}_t = (1 - t)\underline{\mathbf{p}}_1 + t\underline{\mathbf{p}}_2,$$
$$\underline{\mathbf{a}}_s = (1 - s)\underline{\mathbf{p}}_1 + s\underline{\mathbf{p}}_2,$$
$$\underline{\mathbf{b}}_t = (1 - t)\underline{\mathbf{p}}_2 + t\underline{\mathbf{p}}_3,$$
$$\underline{\mathbf{b}}_s = (1 - s)\underline{\mathbf{p}}_2 + s\underline{\mathbf{p}}_3.$$

Let $\underline{\mathbf{c}}$ be the intersection of the straight lines $\underline{f}a_t \wedge \underline{\mathbf{b}}_t$ and $\underline{f}a_s \wedge \underline{\mathbf{b}}_s$. Then $\underline{\mathbf{c}}$ satisfies the two equations

$$\underline{\mathbf{c}} = (1 - s)\underline{\mathbf{a}}_t + s\underline{\mathbf{b}}_t \quad \text{and} \quad \underline{\mathbf{c}} = (1 - t)\underline{\mathbf{a}}_s + t\underline{\mathbf{b}}_s.$$

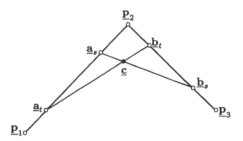

Figure 1.13. Menelaos' theorem: the point $\underline{c}$ may be obtained from linear interpolation at t or at s.

This theorem follows directly from the above definitions. Thus it does not matter if we find $\underline{c}$ by first performing linear interpolation with respect to t and then with respect to s or the other way around.

Notice also that the four collinear points $\underline{p}_1, \underline{a}_t, \underline{a}_s, \underline{p}_2$ as well as the four collinear points $\underline{p}_2, \underline{b}_t, \underline{b}_s, \underline{p}_3$ are projective maps of the four points $0, t, s, 1$ on the real line.

1.9 Duality

Assume we are given two triples of reals, say a, b, c and r, s, t. If we know that their dot product vanishes, i.e., $ar + bs + ct = 0$, how can we interpret this in terms of projective geometry? We could say that the triple a, b, c represents a point, the triple r, s, t represents a line, and the point lies on the line. Just as well, of course, we might assume that a, b, c is the line and r, s, t is a point on it.

Or consider the statement that one triple of reals is the cross product of two other triples. This could mean that a line is being defined by two points, but it might also mean that a point is being defined as the intersection of two lines.

This possibility of dual geometric interpretations of an algebraic statement is fundamental to all of projective geometry, and is called *the principle of duality*. Algebraically, all statements in projective geometry are concerned with certain determinants or dot products vanishing, or certain cross product identities. All those algebraic statements have two geometric interpretations! The principle of duality goes back to J. Poncelet, J. Gergonne, and J. Plücker and was developed around 1830.

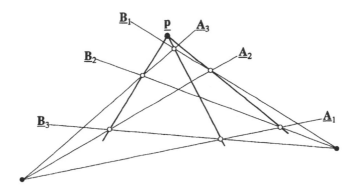

Figure 1.14. The dual Pappus theorem: the indicated (heavy) lines are concurrent.

Instead of elaborating on this theoretically, we will produce a dual interpretation of Pappus' theorem. Stated algebraically, Pappus' theorem says: if $\det[\underline{a}_1, \underline{a}_2, \underline{a}_3] = 0$ and $\det[\underline{b}_1, \underline{b}_2, \underline{b}_3] = 0$, then also $\det[\underline{p}_1, \underline{p}_2, \underline{p}_3] = 0$, with the $\underline{p}_i$ being defined by (1.15).

To obtain the dual statement, we now say: if $\det[\underline{A}_1, \underline{A}_2, \underline{A}_3] = 0$ and $\det[\underline{B}_1, \underline{B}_2, \underline{B}_3] = 0$, then also $\det[\underline{P}_1, \underline{P}_2, \underline{P}_3] = 0$, with $\underline{P}_1 = [\underline{A}_2 \wedge \underline{B}_3] \wedge [\underline{A}_3 \wedge \underline{B}_2]$, etc. Since we already proved Pappus' theorem, this dual version comes for free! Its proof would be a virtual carbon copy of the proof of Pappus' theorem, and can therefore be omitted.

We still need a geometric interpretation of this dual Pappus theorem: it is a statement about two triples of concurrent lines. Intersect corresponding lines as shown in Fig. 1.14, and form lines $\underline{P}_i$ as shown. The dual Pappus theorem states that all three $\underline{P}_i$ are concurrent. The point that they all meet in is called the *Pappus point.*

The principle of duality may appear rather theoretical; yet it will be very useful later in applications such as developable surfaces.

1.10   The Extended Affine Plane

We now return to the discussion from the beginning of this chapter. There we considered the plane $z = 1$ in euclidean three-space and projected 3D lines and planes into it, thus obtaining points and lines of the projective plane $I\!P^2$. Taken simply as a subset of three-space, the plane $z = 1$ is an affine plane. We shall now discuss these two interpretations of the plane — projective and affine.

A point $\mathbf{x} = [x, y]^T$ of the affine plane may be identified with the 3D point $\underline{\mathbf{x}} = [x, y, 1]^T$ in the plane $z = 1$. Interpreted as a projective point, $\underline{\mathbf{x}}$ is also identified with all other points $\underline{\mathbf{x}} \hat{=} [\alpha x, \alpha y, \alpha]^T$ for $\alpha \neq 0$. See Fig. 1.2 for an illustration.

We then have: a projective point $\underline{\mathbf{x}} = [x, y, z]^T$ corresponds to the affine point $\mathbf{x} = [x/z, y/z]^T$. We say that $\underline{\mathbf{x}}$ are the *homogeneous coordinates* of $\mathbf{x}$. For now we must assume $z \neq 0$; we will treat the case $z = 0$ shortly.

What is the affine counterpart $\mathbf{L}$ of a projective line $\underline{\mathbf{L}}$? If $\underline{\mathbf{L}} = [l, m, n]$, then $\mathbf{L} = [l/n, m/n]$. This has to be interpreted as the coefficients of $\mathbf{L}$'s implicit equation $(l/n)x + (m/n)y + 1 = 0$. Again, we must require that $n \neq 0$.

If

$$\underline{\mathbf{x}} = \alpha\underline{\mathbf{a}} + \beta\underline{\mathbf{b}} = \begin{bmatrix} \alpha a_x + \beta b_x \\ \alpha a_y + \beta b_y \\ \alpha a_z + \beta b_z \end{bmatrix}$$

holds for three projective points, what is the corresponding relation for their affine counterparts? We obtain

$$\begin{aligned} \mathbf{x} &= \frac{1}{\alpha a_z + \beta b_z} \begin{bmatrix} \alpha a_x + \beta b_x \\ \alpha a_y + \beta b_y \end{bmatrix} \\ &= \frac{\alpha a_z}{\alpha a_z + \beta b_z} \begin{bmatrix} a_x/a_z \\ a_y/a_z \end{bmatrix} + \frac{\beta b_z}{\alpha a_z + \beta b_z} \begin{bmatrix} b_x/b_z \\ b_y/b_z \end{bmatrix} \\ &= \frac{\alpha a_z}{\alpha a_z + \beta b_z}\mathbf{a} + \frac{\beta b_z}{\alpha a_z + \beta b_z}\mathbf{b}. \end{aligned}$$

Notice how the coefficients of the affine points $\mathbf{a}$ and $\mathbf{b}$ sum to one, thus forming a *barycentric combination,* also called an affine combination. While we can form arbitrary linear combinations of projective points, this is illegal for affine points; the only admissible combination for them is the barycentric combination.

In affine space, we can talk about phenomena that do not exist in projective space. For example: affine lines may be parallel, while projective ones cannot. Take two projective lines $\underline{\mathbf{L}} = [l, m, n]$ and $\underline{\mathbf{M}} = [l, m, k]$. They intersect in the projective point $\underline{\mathbf{L}} \wedge \underline{\mathbf{M}}$, whose third component is zero. Thus it does not have an affine counterpart! We remedy this situation by adding *points at infinity* to the set of affine points. These are all the points with zero third components.

The name "points at infinity" is justified by the informal observation that the intersection of two lines moves further and further towards infinity as the lines become more and more parallel. All affine points at infinity lie on a projective line: it is the line $\underline{\mathbf{L}} = [0, 0, 1]$. Its affine counterpart is the *line at infinity* $\mathbf{L}_\infty$. If we "add" this line to the affine plane, we speak of the extended affine plane $I\!\!E^2$. We may think of this line as the "horizon" of the affine plane. Another word for points at infinity is "ideal points", lying on the "ideal line". The extended affine plane is a model for the projective plane in the same sense as the models that will be discussed in Section 1.11.

The difference $\mathbf{a} - \mathbf{b}$ of two affine points is a *vector*. But the difference $\underline{\mathbf{a}} - \underline{\mathbf{b}}$ of two projective points is yet another projective point; it is on the line through $\underline{\mathbf{a}}$ and $\underline{\mathbf{b}}$. Let $\underline{\mathbf{a}} = [a_x, a_y, 1]^{\mathrm{T}}$ and $\underline{\mathbf{b}} = [b_x, b_y, 1]^{\mathrm{T}}$ be the projective versions of two affine points $\mathbf{a}$ and $\mathbf{b}$. The difference of the affine points is an affine vector; the difference of the projective points corresponds to a point at infinity. Thus affine points at infinity correspond to vectors!

The affine vector $\mathbf{x}$ corresponds to a point at infinity, but so does $\alpha\mathbf{x}$ for any $\alpha \neq 0$. Thus a point at infinity does not only correspond to a single vector, but to all vectors "pointing" to it.

1.11   Topological Aspects

In our investigation of the projective plane, we have so far pointed out *geometric* features, focusing on properties such as collinearity, duality, etc. There is another aspect that we have not addressed yet: the projective plane has a different *topology* than the affine plane. The realization of this fact led to some investigations by computer graphics researchers, since homogeneous coordinates (and hence projective geometry) are one of the standard tools in that field. See Riesenfeld [130] for an introduction.

Let $\underline{\mathbf{L}}, \underline{\mathbf{M}}, \underline{\mathbf{N}}$ be three nonconcurrent lines in the projective plane. They divide $I\!\!P^2$ into four regions: let $\underline{\mathbf{a}}, \underline{\mathbf{b}}, \underline{\mathbf{c}}$ be their intersections, and also pick a fourth point $\underline{\mathbf{d}}$ not on any of the lines. These four points form a projective reference frame, as illustrated by Fig. 1.8. We can identify the following four regions:

$$\mathcal{R}_1 : \begin{bmatrix} + \\ + \\ + \end{bmatrix}, \quad \mathcal{R}_2 : \begin{bmatrix} - \\ + \\ + \end{bmatrix}, \quad \mathcal{R}_3 : \begin{bmatrix} + \\ - \\ + \end{bmatrix}, \quad \mathcal{R}_4 : \begin{bmatrix} + \\ + \\ - \end{bmatrix},$$

where "+" stands for points with positive projective coordinates, and "-" for those with negative ones. Fig. 1.15 illustrates.

This property may be utilized to derive what is known as the *circular model* of $I\!\!P^2$. It is achieved by a simple procedure: first, take the "upper" half of the unit sphere (the semisphere whose points have nonnegative z-coordinates). Project an arbitrary 3D point onto this semisphere, and then use a parallel projection to map it to a point inside the unit circle

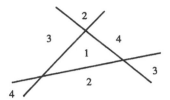

Figure 1.15. A division of the projective plane: three lines divide it into four regions.

of the $x, y-$ plane. Now we can say that every point inside the unit circle corresponds to a point in the projective plane, those points being identical to rays from the origin. A line in $I\!\!P^2$ corresponds to a 3D plane through the origin; it is projected onto a great semicircle and further onto an ellipse inside the unit circle. See Fig. 1.16 for an illustration.

The perimeter of the unit circle, marked by a gray shade in the figure, corresponds to the points at infinity of the extended affine plane. Any two antipodal points are treated as being identical. The unit circle itself then corresponds to the line at infinity.

In $I\!\!P^2$, there are no parallel lines; but in the extended affine plane, they may be defined: two lines are parallel if their intersection is on the line at infinity. Fig. 1.17 shows the circular model of two families of parallel lines.

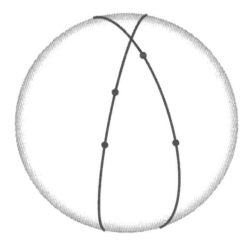

Figure 1.16. The circular model of the projective plane: the unit disc models the projective plane. Some lines and points are shown.

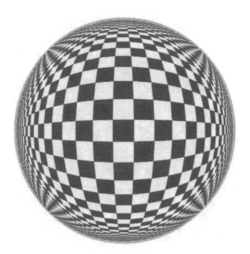

Figure 1.17. The circular model of the projective plane: two families of parallel lines.

In order to make two antipodal points on the unit circle coincide in a "physical" model of $I\!P^2$, we could lift our circular disc out of the plane, deform it and "glue" corresponding antipodal points together. Since we must do this for *all* point pairs of the unit circle, we will obtain a self-intersecting surface. The book by Hilbert and Cohn-Vossen [86] gives a beautiful illustration.

A different way of deforming the circular model results in *Boy's surface*. A steel model of this intriguing surface has been erected[8] in front of the Mathematics Research Center in Oberwolfach, Germany.

A completely different model of the projective plane is the *spherical model*. A point of the projective plane is identified with a pair of antipodal points on the unit sphere. Lines in the projective plane are identified with great circles on the sphere. Any two distinct great circles define a point, and any two distinct points define a great circle.

This model, illustrated by Fig. 1.18, shows best that the projective plane has a different topology than the affine plane. Any straight line divides the affine plane into two separate half planes. A point in one half plane can only be connected to one in the other half plane by crossing the line. But a line in the projective plane does not divide it: a great circle does not divide our spherical model. Every point, being identified with its antipodal point, is at the same time on either side of the great circle!

---

[8]The structure was designed and manufactured by Mercedes-Benz, using a NURB approximation.

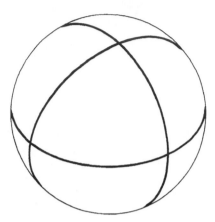

Figure 1.18. The spherical model of the projective plane: projective points are identified with antipodal point pairs; projective lines are identified with great circles.

Thus the notion of a half plane is meaningless in projective geometry. Similarly, it is meaningless to talk about a point being "inside" or "outside" a given triangle.[9] It is possible to modify the classical definition of the projective plane so as to allow for the concepts of half planes, or, more generally, of *orientation*. In essence, oriented projective geometry only treats coordinates as being equivalent if they differ by a *positive* factor; see J. Stolfi [143].

1.12   Parametric Lines

In classical projective geometry, one uses the term "range" when referring to the set of points on a line. We will use the term *parametric line* for a line $\underline{\mathbf{L}}$ that is defined by two points $\underline{\mathbf{a}}$ and $\underline{\mathbf{b}}$: any point on that line may be written as

$$\underline{\mathbf{x}}(t) = (1 - t)\underline{\mathbf{a}} + t\underline{\mathbf{b}}; \quad t \in \mathbb{R}. \tag{1.16}$$

Eq. (1.16) is *linear* in the parameter t; it is a very simple instance of a *parametric curve*. The process of obtaining $\underline{\mathbf{x}}(t)$ is called *linear interpolation*

---

[9]This can cause problems in computer graphics when clipping is performed in homogeneous coordinates, see Blinn/Newell [20].

Any function of a parameter t may be differentiated: we call the derivative $\underline{\dot{x}}(t)$ and find it in a very straightforward way:

$$\underline{\dot{x}}(t) = \underline{b} - \underline{a}. \tag{1.17}$$

This derivative is itself a point on $\underline{L}$! Clearly any linear combination of $\underline{a}$ and $\underline{b}$ defines a point on $\underline{L}$, and $\underline{b} - \underline{a}$ is just one instance of a linear combination.

If we multiply the coordinates of one of our points by some scalar, that point will not change position, and neither will $\underline{L}$ — but the manner in which the parameter t traverses it will change. If we replace the coordinates of $\underline{b}$ by $c\underline{b}$, we call the generic point on the line $\underline{y}(t)$, and its derivative becomes

$$\underline{\dot{y}}(t) = c\underline{b} - \underline{a}. \tag{1.18}$$

This is yet another point on $\underline{L}$, but it is not $\underline{\dot{x}}(t)$! Thus we see that the same line may be described by a variety of parametric equations.

The derivative of a parametric line in $I\!\!P^2$ is a point; what happens if we project the line into affine space? Equation (1.16) has the following affine counterpart:

$$\mathbf{x}(t) = \frac{(1-t)a_z\mathbf{a} + tb_z\mathbf{b}}{(1-t)a_z + tb_z}; \quad t \in I\!\!R. \tag{1.19}$$

If we want to differentiate this, the quotient rule is called for and yields:

$$\dot{\mathbf{x}}(t) = \frac{a_z b_z [\mathbf{b} - \mathbf{a}]}{(1-t)a_z + tb_z}. \tag{1.20}$$

Now $\dot{\mathbf{x}}(t)$ is a vector, just as expected.

1.13   Lines and Line Segments

In euclidean geometry, we use not only lines, but *line segments*. This causes a problem in projective geometry. Consider Fig. 1.19. The left one looks "normal," but not the right one: it crosses the line at infinity. Classical projective geometry would not give any preference to either of the two "segments."

Here is how we will distinguish the two cases: borrowing from oriented projective geometry, we define the line segment connecting two points $\underline{a}$ and $\underline{b}$ to be the set of all points $\underline{x} = (1-t)\underline{a} + t\underline{b}$ where $t \in [0,1]$ *and* the $z-$components of $\underline{a}$ and $\underline{b}$ are both positive. With this definition, we are departing somewhat from strict projective geometry — in all "real life" cases, the sign condition will be met, and we should not expect any complications.

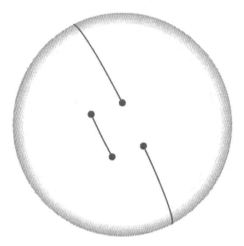

Figure 1.19. Line segments: both ways of connecting two points are considered equivalent in projective geometry.

We may now define a *polygon* connecting points $\underline{\mathbf{a}}_0, \ldots, \underline{\mathbf{a}}_n$: it is the set of all line segments connecting subsequent $\underline{\mathbf{a}}_i$. When referring to a polygon, we typically assume that the z−components of all points are positive. For it they are not, we get "polygons" like the one in Fig. 1.20!

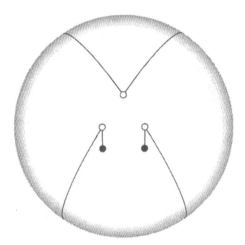

Figure 1.20. Polygons: the third point has a negative z−component, resulting in two crossings of the line at infinity.

1.14 Projective Three-space

All projective concepts so far were restricted to the plane. Once the planar case is understood, it is not hard to explore higher-dimensional spaces, in particular projective three-space. Luckily, no unexpected phenomena arise. We now deal with points, lines, and planes as the basic entities. Any two planes intersect in a line, a line and a plane intersect in a point (or the line is contained in the plane), and two lines may intersect (then they are *coplanar*) or not (then they are *skew*). Planes and points are dual to each other, while lines are dual to themselves. If one were to pursue three-dimensional projective geometry in detail, a different set of symbols would be necessary (involving either tensors or so-called Pluecker coordinates).

A point $\underline{p} \in I\!\!P^3$ is expressed in terms of four coordinates, not all of them zero:

$$\underline{p} = \begin{bmatrix} x \\ y \\ z \\ w \end{bmatrix}.$$

Our previous intuitive development interpreted $I\!\!P^2$ as the projection of $I\!\!E^3$ into the plane $z = 1$ through the origin. We may now think of $I\!\!P^3$ as the projection of $I\!\!E^4$ into the (three-dimensional) hyperplane $w = 1$ through the origin.

A plane is also represented by a quadruple of reals, written as a row:

$$\underline{P} = [k, l, m, n].$$

A point $\underline{p}$ lies on a plane $\underline{P}$ if $\underline{P}\underline{p} = 0$.

Three points $\underline{P}_1, \underline{P}_2, \underline{P}_3$ determine a plane $\underline{P}$:

$$\underline{P} = [\ \det P_1, -\det P_2, \det P_3, -\det P_4\ ], \qquad (1.21)$$

where P_i is obtained by deleting the ith row from the matrix

$$\begin{bmatrix} x_1 & x_2 & x_3 \\ y_1 & y_2 & y_3 \\ z_1 & z_2 & z_3 \\ w_1 & w_2 & w_3 \end{bmatrix},$$

We will use the notation

$$\underline{P} = <\underline{P}_1, \underline{P}_2, \underline{P}_3>.$$

In the same manner, three planes $\underline{P}_1, \underline{P}_2, \underline{P}_3$ define the point $\underline{p}$ in which they intersect:

$$\underline{p} = <\underline{P}_1, \underline{P}_2, \underline{P}_3>.$$

The line obtained by connecting two points or by intersecting two planes is not so easily described — here, the concept of *Plücker* coordinates is used, but we shall not describe it here. See [144] for details.

1.15   Problems

1. Equation (1.2) shows how to compute the line through two distinct points. Draw a figure like Fig. 1.4 to illustrate the idea.

2. After completing Problem 1, illustrate (1.4).

3. Use the spherical model of the projective plane to model the extended affine plane; including the definition of parallel lines.

4. What are the coordinates of the fundamental line from Section 1.5?

5. What is the dual of Fig. 1.9?

6. Four points are given as

$$
\underline{a} = \begin{bmatrix} 0 \\ 0 \\ 1 \end{bmatrix}, \underline{b} = \begin{bmatrix} 1 \\ 0 \\ 1 \end{bmatrix}, \underline{c} = \begin{bmatrix} 1 \\ 2 \\ 2 \end{bmatrix}, \underline{d} = \begin{bmatrix} 1 \\ 1 \\ 2 \end{bmatrix}.
$$

Rescale all points such that the configuration of Fig. 1.8 is achieved. (Note: part of this configuration is shown in Fig. 1.7.)

2

Projective Maps

Geometries are more than just a set of objects, like lines and points. A geometry is also characterized by properties of these objects under the maps of the geometry. Thus Euclidean geometry is characterized by Euclidean maps, which leave lengths and angles unchanged. Affine geometry is characterized by affine maps, which leave the ratio of three collinear points unchanged. Finally, projective geometry is characterized by projective maps which leave the *cross ratio* of four collinear points unchanged. In this chapter, we shall discuss those aspects of projective maps that will later be relevant for NURBS.

2.1 Perspectivities

We have mentioned projections of three-space onto a plane before. At a lower level, one might try to project from a line onto another line, as illustrated in Fig. 2.1. Let two points $\underline{p}_0, \underline{p}_1$ be given on a line $\underline{L}$ and two points $\underline{q}_0, \underline{q}_1$ on another line $\underline{M}$. If $\underline{c}$ is a point on neither $\underline{L}$ nor $\underline{M}$, then we may define a projection of $\underline{L}$ onto $\underline{M}$ through $\underline{c}$ in the following way: if $\underline{x}$ is on $\underline{L}$, then its image $\underline{y}$ on $\underline{M}$ under a projection Φ is defined by

$$\underline{y} = \Phi(\underline{x}) = [\underline{c} \wedge \underline{x}] \wedge \underline{M}. \tag{2.1}$$

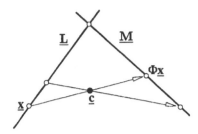

Figure 2.1. Perspectivities: $\underline{\mathbf{x}} = \Phi\underline{\mathbf{x}}$ is defined as a projection of $\underline{\mathbf{x}}$ through $\underline{\mathbf{c}}$ onto $\underline{\mathbf{M}}$.

Such maps are called *perspectivities*. The point $\underline{\mathbf{c}}$ is called the *center* of the perspectivity. We see immediately that the inverse map from $\underline{\mathbf{M}}$ onto $\underline{\mathbf{L}}$ is defined in the same way, making Φ a bijection.[1]

Let $\underline{\mathbf{a}}$ be the intersection of $\underline{\mathbf{L}}$ and $\underline{\mathbf{M}}$. Then $\Phi\underline{\mathbf{a}} = \underline{\mathbf{a}}$, making $\underline{\mathbf{a}}$ a *fixed point* of the perspectivity. Let

$$\underline{\mathbf{x}} = u\underline{\mathbf{p}}_0 + v\underline{\mathbf{p}}_1$$

and

$$\underline{\mathbf{y}} = \hat{u}\underline{\mathbf{q}}_0 + \hat{v}\underline{\mathbf{q}}_1,$$

Since the points $\underline{\mathbf{c}}, \underline{\mathbf{x}}, \underline{\mathbf{y}}$ are collinear by construction, we have that

$$\det[\underline{\mathbf{c}}, u\underline{\mathbf{p}}_0 + v\underline{\mathbf{p}}_1, \hat{u}\underline{\mathbf{q}}_0 + \hat{v}\underline{\mathbf{q}}_1] = 0.$$

Using the multilinearity property of determinants, we may rewrite this as

$$[u, v] \begin{bmatrix} m_{00} & m_{01} \\ m_{10} & m_{11} \end{bmatrix} \begin{bmatrix} \hat{u} \\ \hat{v} \end{bmatrix} = 0, \tag{2.2}$$

where we have used the abbreviations

$$m_{ij} = \det[\underline{\mathbf{c}}, \underline{\mathbf{p}}_i, \underline{\mathbf{q}}_j].$$

Equation (2.2) relates projective coordinates $\mathbf{u} = [u, v]^{\mathrm{T}}$ on the pre-image line to the image coordinates $\hat{\mathbf{u}} = [\hat{u}, \hat{v}]^{\mathrm{T}}$, the image being defined by a perspectivity. Setting $M = \{m_{ij}\}$, we may shorten (2.2) to

$$\mathbf{u}^{\mathrm{T}} M \hat{\mathbf{u}} = 0. \tag{2.3}$$

We note a special case: if $\underline{\mathbf{q}}_0 = \Phi\underline{\mathbf{p}}_0$ and $\underline{\mathbf{q}}_1 = \Phi\underline{\mathbf{p}}_1$, then M's diagonal entries are zero.

---

[1]The reader is invited to work out the dual concept: mapping a pencil onto another pencil!

2.2   Cross Ratios

As mentioned in the introduction to this chapter, cross ratios play a central role in projective geometry. They are defined as follows: Let $\underline{\mathbf{L}}$ be a line and $\underline{\mathbf{a}}$ and $\underline{\mathbf{b}}$ two points on it. If two additional points are defined by

$$\underline{\mathbf{x}} = u_1 \underline{\mathbf{a}} + v_1 \underline{\mathbf{b}} \quad \text{and} \quad \underline{\mathbf{y}} = u_2 \underline{\mathbf{a}} + v_2 \underline{\mathbf{b}},$$

then their *cross ratio* cr is given by

$$\text{cr}(\underline{\mathbf{a}}, \underline{\mathbf{x}}, \underline{\mathbf{y}}, \underline{\mathbf{b}}) = \frac{v_1}{u_1} \frac{u_2}{v_2}. \tag{2.4}$$

This particular definition will be useful later; it is not, however, what one finds in classical texts on projective geometry. The commonly used classical definition $\text{cr}_{\text{classic}}$ is related to ours by

$$\text{cr}(\underline{\mathbf{a}}, \underline{\mathbf{x}}, \underline{\mathbf{y}}, \underline{\mathbf{b}}) = \text{cr}_{\text{classic}}(\underline{\mathbf{x}}, \underline{\mathbf{a}}, \underline{\mathbf{y}}, \underline{\mathbf{b}}),$$

and so both definitions can be viewed as being equivalent.

Note that we may multiply the coordinates of any point by an arbitrary nonzero number, and the cross ratio will not change! Also note that $\text{cr}(\underline{\mathbf{a}}, \underline{\mathbf{x}}, \underline{\mathbf{y}}, \underline{\mathbf{b}}) = \text{cr}(\underline{\mathbf{b}}, \underline{\mathbf{y}}, \underline{\mathbf{x}}, \underline{\mathbf{a}})$.

Now let $\underline{\mathbf{c}}$ be a point not on $\underline{\mathbf{L}}$, and define

$$
\begin{aligned}
\underline{\mathbf{A}} &= \underline{\mathbf{a}} \wedge \underline{\mathbf{c}}, \\
\underline{\mathbf{X}} &= \underline{\mathbf{x}} \wedge \underline{\mathbf{c}}, \\
\underline{\mathbf{Y}} &= \underline{\mathbf{y}} \wedge \underline{\mathbf{c}}, \\
\underline{\mathbf{B}} &= \underline{\mathbf{b}} \wedge \underline{\mathbf{c}}.
\end{aligned}
$$

Since $\underline{\mathbf{X}} = u_1 \underline{\mathbf{A}} + v_1 \underline{\mathbf{B}}$ and $\underline{\mathbf{Y}} = u_2 \underline{\mathbf{A}} + v_2 \underline{\mathbf{B}}$ because of (1.14), we can adopt the same definition for the cross ratio of four concurrent lines:

$$\text{cr}(\underline{\mathbf{A}}, \underline{\mathbf{X}}, \underline{\mathbf{Y}}, \underline{\mathbf{B}}) = \text{cr}(\underline{\mathbf{a}}, \underline{\mathbf{x}}, \underline{\mathbf{y}}, \underline{\mathbf{b}}). \tag{2.5}$$

See Fig. 2.2 for an illustration.

We may now state that *perspectivities leave cross ratios invariant*. For in Fig. 2.2, we may draw any other line $\underline{\mathbf{M}}$. It will intersect $\underline{\mathbf{A}}, \underline{\mathbf{X}}, \underline{\mathbf{Y}}, \underline{\mathbf{B}}$ in four points — and those four points, being related to $\underline{\mathbf{a}}, \underline{\mathbf{x}}, \underline{\mathbf{y}}, \underline{\mathbf{b}}$ by a perspectivity, will be also in the same cross ratio as $\underline{\mathbf{A}}, \underline{\mathbf{X}}, \underline{\mathbf{Y}}, \underline{\mathbf{B}}$ and hence as $\underline{\mathbf{a}}, \underline{\mathbf{x}}, \underline{\mathbf{y}}, \underline{\mathbf{b}}$.

A perspectivity from a line $\underline{\mathbf{L}}$ to another line $\underline{\mathbf{M}}$ is determined by two pre-image points on $\underline{\mathbf{L}}$ and two image points on $\underline{\mathbf{M}}$, since this is enough information to determine the center $\underline{\mathbf{c}}$. Four points with cross ratio 1/2 are

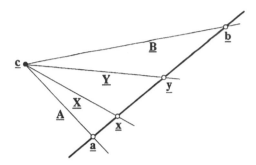

Figure 2.2. Cross ratios: the four collinear points have the same cross ratio as the four concurrent lines.

called *harmonic.*[2] We sometimes relax this definition and label points as harmonic if some permutation of them yields the cross ratio $1/2$. Harmonic points will play an important role in the projective theory of conic sections. If $\underline{a}$ and $\underline{b}$ are two distinct points, then

$$cr(\underline{a}, \underline{a} + \underline{b}, \underline{b}, \underline{b} - \underline{a}) = \frac{1}{2}. \tag{2.6}$$

We see this by writing

$$\underline{a} + \underline{b} = 2 \cdot \underline{a} + 1 \cdot (\underline{b} - \underline{a}),$$
$$\underline{b} = 1 \cdot \underline{a} + 1 \cdot (\underline{b} - \underline{a}),$$

from which our claim follows immediately. In an affine setting, this amounts to stating that two points, their midpoint, and their difference vector are harmonic.

   Another set of harmonic points is found by adding a line and a point to Fig. 1.8, as shown in Fig. 2.3. If we draw the line through $\underline{a} + \underline{c}$ and $\underline{b} + \underline{c}$ and intersect it with $\underline{a} \wedge \underline{b}$, the resulting point is $\underline{a} - \underline{b}$, which is, of course, equivalent to $\underline{b} - \underline{a}$. Again, we have four harmonic points on the line $\underline{a} \wedge \underline{b}$.

   We could have started this construction by interpreting the four points $\underline{d}, \underline{b} + \underline{c}, \underline{c}$, and $\underline{a} + \underline{c}$ in Fig. 2.3 as our initial quadrilateral. Thus four harmonic points are obtained as follows: take any quadrilateral, and intersect opposite edges, resulting in two points. Intersect the line formed by these two points with the two diagonals of the quadrilateral, and we have constructed four harmonic points. A quadrilateral with both diagonals added to it is called a *complete quadrilateral* . Thus complete quadrilaterals are intimately connected with harmonic points.

---

[2]In the classical definition of cross ratios, they are called harmonic if their "classic cross ratio" $cr_{classic}$ equals -1.

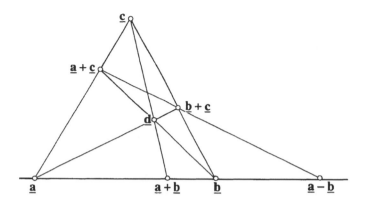

Figure 2.3. Harmonic points: the point $\underline{\mathbf{d}}$ generates four harmonic points on the line $\underline{\mathbf{a}} \wedge \underline{\mathbf{b}}$.

If three points $\underline{\mathbf{a}}, \underline{\mathbf{b}}, \underline{\mathbf{c}}$ are given on a line, we may ask for the point $\underline{\mathbf{x}}$ that satisfies $\mathrm{cr}(\underline{\mathbf{a}}, \underline{\mathbf{x}}, \underline{\mathbf{b}}, \underline{\mathbf{c}}) = \rho$ for some given cross ratio ρ. This point is uniquely defined. To compute it, assume that $\underline{\mathbf{b}} = u\underline{\mathbf{a}} + v\underline{\mathbf{c}}$. Let us choose the representation $\underline{\mathbf{x}} = \underline{\mathbf{a}} + t\underline{\mathbf{c}}$ for the point $\underline{\mathbf{x}}$. We thus need to find t; it can be computed from the known cross ratio relation

$$\rho = \frac{u}{v} \cdot t. \tag{2.7}$$

We will later be dealing with points on a line that are simply referenced by real numbers: any point $\underline{\mathbf{p}}$ may be assigned a *parameter* t with respect to some coordinate system on the line. If we have four points $\underline{\mathbf{p}}_1, \underline{\mathbf{p}}_2, \underline{\mathbf{p}}_3, \underline{\mathbf{p}}_4$ on a line and each $\underline{\mathbf{p}}_i$ is assigned a parameter value t_i, then we will use $\mathrm{cr}(t_1, t_2, t_3, t_4)$ to mean the usual $\mathrm{cr}(\underline{\mathbf{p}}_1, \underline{\mathbf{p}}_2, \underline{\mathbf{p}}_3, \underline{\mathbf{p}}_4)$. Conversely, if we are given four real numbers a_1, a_2, a_2, a_4, then we define

$$\mathrm{cr}(a_1, a_2, a_2, a_4) = \mathrm{cr}\left(\frac{a_1}{1}, \frac{a_2}{1}, \frac{a_3}{1}, \frac{a_4}{1} \right), \tag{2.8}$$

i.e., we associate each a_i with a point whose coordinates with respect to some coordinate system on the line are given by a_i and 1.

2.3   Projectivities

Perspectivities are only a subset of a more general type of map from one line to another. Their sole characterization is that they leave cross ratios unchanged. Such maps are called *projectivities*.

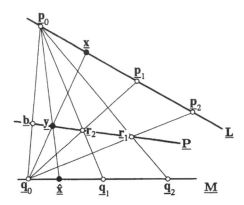

Figure 2.4. Projectivities: $\underline{x}$'s image $\hat{\underline{x}}$ may be obtained from two perspectivities.

We will now proceed to give a geometric construction for a projectivity.

Let $\underline{p}_0, \underline{p}_1, \underline{p}_2$ be three points on a line $\underline{L}$. On a second line $\underline{M}$, we prescribe three points $\underline{q}_0, \underline{q}_1, \underline{q}_2$ to be the images of the corresponding $\underline{p}_i$. Our aim is this: for any point $\underline{x}$ on $\underline{L}$, find an image point $\hat{\underline{x}}$ on $\underline{M}$ such that

$$\mathrm{cr}(\underline{p}_0, \underline{x}, \underline{p}_1, \underline{p}_2) = \mathrm{cr}(\underline{q}_0, \hat{\underline{x}}, \underline{q}_1, \underline{q}_2).$$

See Fig. 2.4 for an illustration. If we are able to produce an $\hat{\underline{x}}$ for any $\underline{x}$, we have defined a projectivity.

We proceed as follows: first, we recall that Pappus' theorem (Section 1.7) asserts the existence of the line $\underline{P}$, namely the Pappus line, with points $\underline{r}_1$ and $\underline{r}_2$ on it as shown in Fig. 2.4.

Next, we define a perspectivity from $\underline{L}$ to $\underline{P}$ with center $\underline{q}_0$. It will map $\underline{x}$ to a point $\underline{y}$ on $\underline{P}$. We observe that

$$\mathrm{cr}(\underline{p}_0, \underline{x}, \underline{p}_1, \underline{p}_2) = \mathrm{cr}(\underline{b}, \underline{y}, \underline{r}_2, \underline{r}_1),$$

with $\underline{b} = [\underline{p}_0 \wedge \underline{q}_0] \wedge \underline{P}$. Then we define another perspectivity; with center $\underline{p}_0$ and mapping $\underline{P}$ to $\underline{M}$. It maps $\underline{y}$ to the desired point $\hat{\underline{x}}$ on $\underline{M}$. We have

$$\mathrm{cr}(\underline{b}, \underline{y}, \underline{r}_2, \underline{r}_1) = \mathrm{cr}(\underline{q}_0, \hat{\underline{x}}, \underline{q}_1, \underline{q}_2),$$

thus asserting that we have in fact found a projectivity from $\underline{L}$ to $\underline{M}$.

In fact, we have not only shown that a projectivity from one line to another is defined by two triples of images and pre-images, but we have also shown that each projectivity may be composed as the concatenation of *two perspectivities*. If the image and pre-image lines are identical, we will need three perspectivities.

Our approach to projectivities was a purely constructive one: given three images and pre-images, we then constructed an image point for any given pre-image point. A more analytical description of projectivities is also available; we will now show that every projectivity can be written in matrix form:

$$\hat{\underline{u}} = \Phi(\underline{u}) = M\underline{u}, \tag{2.9}$$

with $\underline{u}$ and $\hat{\underline{u}}$ projective coordinates on the lines $\underline{L}$ and $\underline{M}$, respectively. (We are using the notation from Section 2.1 here.)

All we have to show is that the map Φ preserves cross ratios. Let $\underline{P}_1 = u_1\underline{P}_0 + v_1\underline{P}_2$ and $\underline{x} = u_2\underline{P}_0 + v_2\underline{P}_2$. Then

$$\begin{aligned}
\hat{\underline{x}} &= M(u_2\underline{P}_0 + v_2\underline{P}_2) \\
&= u_2 M\underline{P}_0 + v_2 M\underline{P}_2 \\
&= u_2\underline{q}_0 + v_2\underline{q}_2.
\end{aligned}$$

The same holds for $\underline{P}_1$ — thus M leaves linear relationships unchanged, and our claim is proved.

2.4   Moebius Transformations

A projectivity can be described by a linear transformation of projective coordinates of the image and pre-image line. We arrive at a different formulation if we endow each line with only *one* parameter. So let the pre-image point $\underline{x}$ be given by

$$\underline{x} = (1-t)\underline{P}_0 + t\underline{P}_2$$

and the image $\hat{\underline{x}}$ by

$$\hat{\underline{x}} = (1-\hat{t})\hat{\underline{P}}_0 + \hat{t}\hat{\underline{P}}_2.$$

If $\underline{P}_1 = (1-u)\underline{P}_0 + u\underline{P}_2$ and $\hat{\underline{P}}_1 = (1-\hat{u})\hat{\underline{P}}_0 + \hat{u}\hat{\underline{P}}_2$, we can define

$$r = \frac{u}{1-u} \quad \text{and} \quad \hat{r} = \frac{\hat{u}}{1-\hat{u}}.$$

The condition of equal cross ratios of image and pre-image points now yields

$$r\frac{1-t}{t} = \hat{r}\frac{1-\hat{t}}{\hat{t}},$$

from which we obtain

$$\hat{t} = \frac{\hat{r}t}{(1-t)r + t\hat{r}}. \tag{2.10}$$

Figure 2.5. Collineations: a checkerboard and some images under collineations. The circular model of $I\!\!P^2$ is used.

This is the desired transformation from t to $\hat{t}$. It is a rational linear transformation, and is often called a *Moebius transformation*.[3] Note that if $r = \hat{r}$, we obtain $\hat{t} = t$.

2.5   Collineations

We have so far studied maps from a line to another line. Going a step further, we may map all of the projective plane $I\!\!P^2$ onto itself, by maps that are called *collineations*. An example is shown in Fig. 2.5: there, a quadrilateral is mapped twice by collineations.

A collineation Φ takes points to points and lines to lines. Moreover, it takes collinear points to collinear points and concurrent lines to concurrent lines. With these definitions, we may already conclude that Φ has a matrix expression when applied to points:

$$\Phi\underline{x} = M\underline{x}, \qquad (2.11)$$

where M is a nonsingular 3×3 matrix. For if three points $\underline{x}, \underline{y}, \underline{z}$, are collinear, we have

---

[3]The terms rational linear transformation or birational transformation are also used.

$$\det[\underline{x}, \underline{y}, \underline{z}] = 0,$$
$$\det[\Phi\underline{x}, \Phi\underline{y}, \Phi\underline{z}] = \det[M\underline{x}, M\underline{y}, M\underline{z}]$$
$$= \det M \det[\underline{x}, \underline{y}, \underline{z}]$$
$$= 0.$$

We can also apply Φ to lines, but now, using the matrix M from (2.11), it takes the form

$$\Phi\underline{L} = \underline{L}M^{-1}. \tag{2.12}$$

This is seen by observing that if $\underline{x}$ lies on $\underline{L}$, then $\Phi\underline{x}$ lies on $\Phi\underline{L}$. In formulas: if $\underline{L}\underline{x} = 0$, then $\Phi\underline{L}\Phi\underline{x} = 0$. We would like to find a matrix N such that $\Phi\underline{L} = \underline{L}N$. Then

$$\Phi\underline{L}\Phi\underline{x} = \underline{L}NM\underline{x} = \underline{L}\underline{x} = 0. \tag{2.13}$$

Let two points be given by

$$\underline{P}_0 = \begin{bmatrix} 0 \\ 1 \\ 1 \end{bmatrix} \quad \text{and} \quad \underline{P}_1 = \begin{bmatrix} 1 \\ 0 \\ 1 \end{bmatrix}.$$

Let a collineation Φ be defined by

$$M = \begin{bmatrix} 1 & 0 & 1 \\ 0 & 1 & 0 \\ 0 & 0 & 1 \end{bmatrix},$$

such that

$$\Phi\underline{P}_0 = \begin{bmatrix} 1 \\ 1 \\ 1 \end{bmatrix} \quad \text{and} \quad \Phi\underline{P}_1 = \begin{bmatrix} 2 \\ 0 \\ 1 \end{bmatrix}.$$

The line through $\underline{p}_0$ and $\underline{p}_1$ is $\underline{L} = [1, 1, -1]$, and the line through $\Phi\underline{p}_0, \Phi\underline{p}_1$ is $\Phi\underline{L} = [1, 1, -2]$. Since

$$M^{-1} = \begin{bmatrix} 1 & 0 & -1 \\ 0 & 1 & 0 \\ 0 & 0 & 1 \end{bmatrix},$$

we have that $\Phi\underline{L} = \underline{L}M^{-1}$.

Example 2.1. Collineations of points and lines.

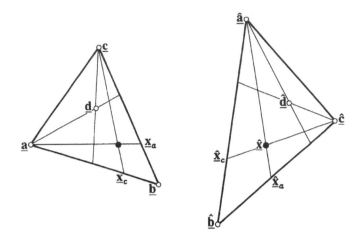

Figure 2.6. Collineations: the image $\hat{\underline{x}}$ of $\underline{x}$ may be found by two cross ratio constructions.

It follows that $NM = I$, the identity matrix, and thus $N = M^{-1}$. Of course any nonzero multiple αM or αM^{-1} could also serve as the desired maps. Example 2.1 illustrates.

Collineations leave cross ratios invariant: this follows easily from the linearity properties of matrix multiplication, in the same way we proved (2.9). Thus if a collineation Φ maps a line $\underline{\mathbf{L}}$ to another line $\Phi\underline{\mathbf{L}}$, these two lines are related by a projectivity.

How many image and pre-image point pairs determine a collineation? Since a collineation is determined by a 3×3 matrix, one might guess that three such pairs are sufficient. But the matrix is only determined up to a constant, and the correct answer is four. We give a geometric demonstration first. Then we will show how to find the matrix that is associated with the collineation.

Referring to Fig. 2.6, we assume that we are given four points $\underline{\mathbf{a}}, \underline{\mathbf{b}}, \underline{\mathbf{c}}, \underline{\mathbf{d}}$ and their images. For a given point $\underline{\mathbf{x}}$, we would like to find its image $\hat{\underline{\mathbf{x}}}$.

First, we construct $\underline{\mathbf{d}}_a = [\underline{\mathbf{a}} \wedge \underline{\mathbf{d}}] \wedge [\underline{\mathbf{b}} \wedge \underline{\mathbf{c}}]$. We find $\hat{\underline{\mathbf{d}}}_a$ from the analogous construction with the image points. Next, we construct $\underline{\mathbf{x}}_a = [\underline{\mathbf{a}} \wedge \underline{\mathbf{x}}] \wedge [\underline{\mathbf{b}} \wedge \underline{\mathbf{c}}]$. Its image point $\hat{\underline{\mathbf{x}}}_a$ is found from the condition that

$$\mathrm{cr}(\underline{\mathbf{b}}, \underline{\mathbf{d}}_a, \underline{\mathbf{x}}_a, \underline{\mathbf{c}}) = \mathrm{cr}(\hat{\underline{\mathbf{b}}}, \hat{\underline{\mathbf{d}}}_a, \hat{\underline{\mathbf{x}}}_a, \hat{\underline{\mathbf{c}}}),$$

by the use of (2.7). In the same manner, we find points $\underline{\mathbf{d}}_c, \hat{\underline{\mathbf{d}}}_c$ and $\underline{\mathbf{x}}_c, \hat{\underline{\mathbf{x}}}_c$. Since $\underline{\mathbf{x}} = [\underline{\mathbf{a}} \wedge \underline{\mathbf{x}}_a] \wedge [\underline{\mathbf{c}} \wedge \underline{\mathbf{x}}_c]$, and since Φ preserves this relationship, we find $\underline{x}$'s image as

$$\hat{\underline{x}} = [\hat{\underline{a}} \wedge \hat{\underline{x}}_a] \wedge [\hat{\underline{c}} \wedge \hat{\underline{x}}_c]. \tag{2.14}$$

Note that $\underline{d}$ does not have to be "inside" the triangle $\underline{a}, \underline{b}, \underline{c}$ — it just should not be on any of its edges.

What is the matrix M that carries four pre-image points to four image points? In order to answer this question, recall that we can always[4] write $\underline{d}$ as $\underline{d} \hat{=} \underline{a} + \underline{b} + \underline{c}$ by an appropriate scaling of coordinates. Thus

$$\underline{d} = \alpha\underline{a} + \beta\underline{b} + \gamma\underline{c}, \tag{2.15}$$

where α, β, γ are those appropriate factors. Equation (2.15) constitutes a three by three linear system for α, β, γ which is solvable. We can now define a projective map of the pre-image plane to itself by means of the matrix $A = [\alpha\underline{a}, \beta\underline{b}, \gamma\underline{c}]^{-1}$.

Similarly, we define a matrix $\hat{A} = [\hat{\alpha}\hat{\underline{a}}, \hat{\beta}\hat{\underline{b}}, \hat{\gamma}\hat{\underline{c}}]^{-1}$ which maps the image plane to itself. The numbers $\hat{\alpha}, \hat{\beta}, \hat{\gamma}$ are the solution of the system

$$\hat{\underline{d}} = \hat{\alpha}\hat{\underline{a}} + \hat{\beta}\hat{\underline{b}} + \hat{\gamma}\hat{\underline{c}}. \tag{2.16}$$

Now the pre-image point $\underline{x}$ has been mapped to the same coordinate triple as its image $\hat{\underline{x}}$:

$$A\underline{x} \hat{=} \hat{A}\hat{\underline{x}}.$$

Consequently, since $\hat{\underline{x}} = M\underline{x}$ holds for all $\underline{x}$, we have

$$A = \hat{A}M$$

and thus

$$M = \hat{A}^{-1}A = [\hat{\alpha}\hat{\underline{a}}, \hat{\beta}\hat{\underline{b}}, \hat{\gamma}\hat{\underline{c}}][\alpha\underline{a}, \beta\underline{b}, \gamma\underline{c}]^{-1}. \tag{2.17}$$

At first sight, it appears as if M depends on our initial ordering of the image and pre-image points. Yet M is uniquely determined (up to a constant factor), but we omit the algebra necessary for a proof of this statement.

Example 2.2 shows an application of this technique: it maps a trapezoid onto a square. The three-dimensional analogue of this particular map is known as the *viewing transformation* in computer graphics, see [73]. There, it maps a truncated pyramid onto a cube.

---

[4]We assume that we are not dealing with degenerate situations here!

Let four pre-image points $\underline{a}, \underline{b}, \underline{c}, \underline{d}$ be given by

$$
\begin{bmatrix} 0 \\ 1 \\ 1 \end{bmatrix}, \begin{bmatrix} 0 \\ -1 \\ 1 \end{bmatrix}, \begin{bmatrix} 1 \\ 2 \\ 1 \end{bmatrix}, \begin{bmatrix} 1 \\ -2 \\ 1 \end{bmatrix},
$$

and their respective images by

$$
\begin{bmatrix} 0 \\ 2 \\ 1 \end{bmatrix}, \begin{bmatrix} 0 \\ -2 \\ 1 \end{bmatrix}, \begin{bmatrix} 1 \\ 2 \\ 1 \end{bmatrix}, \begin{bmatrix} 1 \\ -2 \\ 1 \end{bmatrix}.
$$

The condition $\alpha\underline{a} + \beta\underline{b} + \gamma\underline{c} = \underline{d}$ leads to a linear system with the solution $\alpha = -2, \beta = 2, \gamma = 1$. The inverse of the matrix $[\alpha\underline{a}, \beta\underline{b}, \gamma\underline{c}]$ is

$$
\begin{bmatrix} 0 & 0 & 1 \\ -2 & -2 & 2 \\ -2 & 2 & 1 \end{bmatrix}^{-1} = \frac{1}{4} \begin{bmatrix} 3 & -1 & -1 \\ 1 & -1 & 1 \\ 4 & 0 & 0 \end{bmatrix}.
$$

Similarly, we find $\hat{\alpha} = -1, \hat{\beta} = 1, \hat{\gamma} = 1$. Using (2.17), we find

$$
M = \frac{1}{4} \begin{bmatrix} 4 & 0 & 0 \\ 0 & 4 & 0 \\ 2 & 0 & 2 \end{bmatrix}.
$$

It is easy to check that indeed $M\underline{a} \hat{=} \hat{\underline{a}}$ etc. Also, note that the point $[-1, 0, 1]^T$ is mapped to the point $[-1, 0, 0]^T$ at infinity.

Example 2.2. Constructing collineations.

2.6 Constructing Collineations

We close this section with a special collineation. In Fig. 2.6, assume that $\underline{a}, \underline{b}, \underline{c}$ are mapped to themselves, only $\underline{d}$ is mapped to a different point $\hat{\underline{d}}$, but such that $\underline{d}, \hat{\underline{d}}, \underline{a}$ are collinear, see Fig. 2.7. This is obviously a special kind of collineation; since it will reappear later, we give it its own name: it is called a *central* collineation.

Defining $\underline{A} = \underline{b} \wedge \underline{c}$, we see that all points of $\underline{A}$ are mapped to themselves: we say that $\underline{A}$ is fixed pointwise by the central collineation. Likewise, all lines in the pencil through $\underline{a}$ are mapped to themselves: we say that $\underline{a}$ is fixed linewise. Thus a central collineation is defined by its *center* $\underline{a}$ and by its *axis* $\underline{A}$.

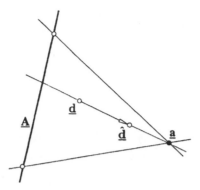

Figure 2.7. Central collineations: these are collineations that are defined by an axis $\underline{\mathbf{A}}$ and a center $\underline{\mathbf{a}}$.

2.7 Affine Maps

In the extended affine plane, we may observe the effect that a projective map has on the line at infinity $\mathbf{L}_\infty$. A projective map that maps $\mathbf{L}_\infty$ to itself is called an *affine map*. Note that not every point on $\mathbf{L}_\infty$ needs to be mapped to itself; all we require is that its image be *somewhere* on $\mathbf{L}_\infty$.

The line at infinity corresponds to the projective points $[x, y, 0]^\mathrm{T}$, which correspond to affine vectors. Since that line is mapped to itself, vectors are mapped to vectors, and so affine points $[x, y, 1]^\mathrm{T}$ must also be mapped to affine points:

$$\Phi : \begin{bmatrix} \star \\ \star \\ 1 \end{bmatrix} \rightarrow \begin{bmatrix} \star \\ \star \\ 1 \end{bmatrix} \quad \text{and} \quad \Phi : \begin{bmatrix} \star \\ \star \\ 0 \end{bmatrix} \rightarrow \begin{bmatrix} \star \\ \star \\ 0 \end{bmatrix}, \quad (2.18)$$

where $\star$ denotes arbitrary real numbers.

These restrictions on Φ imply restrictions on the matrix M: we find that M's last row must be of the form

$$[m_{31}, m_{32}, m_{33}] = [0, 0, 1]$$

or any nonzero multiple of it. We can thus rewrite M as

$$M = \begin{bmatrix} a_{11} & a_{12} & v_1 \\ a_{21} & a_{22} & v_2 \\ 0 & 0 & 1 \end{bmatrix}.$$

With

$$\underline{x} = \begin{bmatrix} x \\ y \\ 1 \end{bmatrix} = \begin{bmatrix} \mathbf{x} \\ 1 \end{bmatrix},$$

we now have

$$M\underline{x} = \begin{bmatrix} a_{11} & a_{12} & v_1 \\ a_{21} & a_{22} & v_2 \\ 0 & 0 & 1 \end{bmatrix} \begin{bmatrix} x \\ y \\ 1 \end{bmatrix} = \begin{bmatrix} A\mathbf{x} + \mathbf{v} \\ 1 \end{bmatrix},$$

with $\mathbf{v} = [v_1, v_2]^{\mathrm{T}}$ and $A = \{a_{ij}\}$. Thus an affine map (of the affine plane to itself) is given by the *point map*

$$\Phi\mathbf{x} = A\mathbf{x} + \mathbf{v}; \quad \mathbf{x} \in I\!\!E^2, \ \mathbf{v} \in I\!\!R^2, \tag{2.19}$$

and the *vector map*

$$\Phi\mathbf{w} = M\mathbf{w}; \quad \mathbf{w} \in I\!\!R^2. \tag{2.20}$$

being obtained in the same way. The vector $\mathbf{v}$ in (2.19) is called the *translational part* of Φ.

A collineation may be defined by the pre-images and images of four given lines. Since affine maps always map one of these lines, namely $\mathbf{L}_\infty$, to itself, an affine map is only determined by the pre-images and images of three given lines. Three lines intersect in three points, and so we may also think of an affine map as being determined by three pre-image/image point pairs.

We pointed out that the fundamental invariant of projective maps is the cross ratio. Affine maps, being special projective maps, leave cross ratios constant, too. But they leave another quantity invariant as well, namely the *ratio* between three collinear points. It is defined as follows: if $\mathbf{b} = (1 - \alpha)\mathbf{a} + \alpha\mathbf{c}$, then

$$\mathrm{ratio}(\mathbf{a}, \mathbf{b}, \mathbf{c}) = \frac{\alpha}{1 - \alpha}. \tag{2.21}$$

As might be expected, ratios are a special case of cross ratios. If $\mathbf{d}$ is a point at infinity, then

$$\mathrm{ratio}(\mathbf{a}, \mathbf{b}, \mathbf{c}) = \mathrm{cr}(\mathbf{a}, \mathbf{b}, \mathbf{c}, \mathbf{d}). \tag{2.22}$$

We will later encounter ratios in the context of rational Bézier curves, see Section 7.1.

2.8   Problems

1. Show that two perspectivities are not sufficient to describe a projectivity of a line onto itself.

2. Let $\underline{L}$ be a line with three points $\underline{a}, \underline{b}, \underline{c}$ on it. Let $\underline{x}$ trace out all of $\underline{L}$. We define four cross ratios at each point on $\underline{L}$:

$$\operatorname{cr}(\underline{x}, \underline{a}, \underline{b}, \underline{c}),$$
$$\operatorname{cr}(\underline{a}, \underline{x}, \underline{b}, \underline{c}),$$
$$\operatorname{cr}(\underline{a}, \underline{b}, \underline{x}, \underline{c}),$$
$$\operatorname{cr}(\underline{a}, \underline{b}, \underline{c}, \underline{x}).$$

Plot four graphs: make $\underline{L}$ the $x-$axis of a standard Euclidean coordinate system and plot the cross ratio values in the y direction.

3. What are the affine analogues of the constructions described by (2.7) and (2.14)?

4. What is the dual of the complete quadrilateral construction from Section 2.2?

5. Draw the dual of Fig. 2.6.

6. If affine maps of points are defined by (2.19), how are affine maps of lines defined?

7. What is the projective pre-image of the affine zero vector?

3

Conics

Conics are the ancestors of all NURBS. By studying conics, we can observe most of the basic NURBS principles.

The classical definition generates conics as the intersections of a cone with a plane. One then obtains the familiar ellipses, hyperbolas, and parabolas. In projective geometry, this distinction is nonexistent, as is demonstrated in Fig. 3.1. There, the tip of the (double) cone is at the origin of three-space, and the cone is intersected with a 3D plane to yield a conic section in three-space. This conic section is projected into the plane $z = 1$, yielding yet another conic section. But we might have intersected our cone with any other plane, yielding a different conic section in three-space, and the projection would again be the same conic in the $z = 1$ plane! This is the reason — informally stated — why in projective geometry, we cannot distinguish between different types of conics.

3.1 Line and Point Conics

We will not use the above approach to conics, but rather follow an approach that is due to J. Steiner:

Figure 3.1. Conic sections: the classical definition intersects a cone with planes.

Let Φ be a projectivity that maps the line $\underline{\mathbf{L}}$ onto another line $\underline{\mathbf{M}}$. We generate a line $\underline{\mathbf{X}}$ by taking a point $\underline{\mathbf{x}}$ on $\underline{\mathbf{L}}$, constructing its image $\hat{\underline{\mathbf{x}}} = \Phi\underline{\mathbf{x}}$, and by connecting the two points:

$$\underline{\mathbf{X}} = \underline{\mathbf{x}} \wedge \hat{\underline{\mathbf{x}}}. \tag{3.1}$$

The collection of all such lines $\underline{\mathbf{X}}$ is called a (line) *conic* Γ. The conic is called *singular* if either $\underline{\mathbf{L}} = \underline{\mathbf{M}}$ or if Φ is a perspectivity. The two lines $\underline{\mathbf{L}}$ and $\underline{\mathbf{M}}$ are called *generating lines*.

If Φ is a perspectivity, the conic degenerates to a point, namely the center of the perspectivity. If the two generating lines coincide, the conic degenerates to that line. If a line $\underline{\mathbf{X}}$ is one of the lines that define Γ, we say that $\underline{\mathbf{X}}$ is on Γ; we also say that $\underline{\mathbf{X}}$ is a *tangent* of Γ, a notion that will be justified below.

Since Φ is a projectivity, it is a bijective map. So we might as well have used Φ^{-1} in our definition. Figures 3.2 and 3.3 illustrate.

A dual definition also exists. It defines a conic as the locus of points, namely all intersections of corresponding lines in two projectively related pencils, see Fig. 3.2. Such conics are called *point conics*, whereas our definition (3.1) would generate *line conics*. Both definitions yield the same class of curves, as we shall see later.

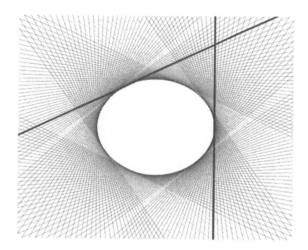

Figure 3.2. Line conics: corresponding points on two lines are connected.

Returning to line conics, it appears that a conic Γ depends crucially on the generating lines $\underline{L}$ and $\underline{M}$. This is not so: any other two lines on it can equally well serve as generating lines.[1] In order to see this, let $\underline{A}$ be a line on the conic with generating lines $\underline{L}$ and $\underline{M}$. Our claim is now that we can also make $\underline{A}$ and $\underline{M}$ generating lines of Γ. Referring to Fig. 3.4, we need projectively related ranges on $\underline{A}$ and $\underline{M}$ such that the connection of corresponding

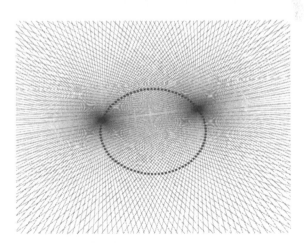

Figure 3.3. Point conics: corresponding lines in two pencils are intersected.

[1]This is a theoretical statement: in a "real" situation, not all generating lines are numerically equivalent!

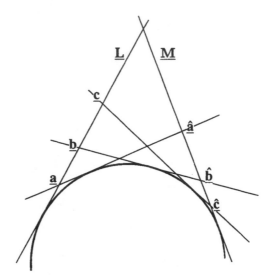

Figure 3.4. Generating lines: any two of the shown lines can be taken as generating lines.

points are lines on Γ. A projectivity from one range to another is determined by three image and pre-image pairs. If we can produce three such pairs, we have proved our claim. We can choose

$$
\begin{array}{ccccccc}
\underline{\mathbf{a}} & \text{on} & \underline{\mathbf{A}} & \text{and} & \underline{\mathbf{L}} \wedge \underline{\mathbf{M}} & \text{on} & \underline{\mathbf{M}}, \\
\underline{\mathbf{B}} \wedge \underline{\mathbf{A}} & \text{on} & \underline{\mathbf{A}} & \text{and} & \hat{\underline{\mathbf{b}}} & \text{on} & \underline{\mathbf{M}}, \\
\underline{\mathbf{C}} \wedge \underline{\mathbf{A}} & \text{on} & \underline{\mathbf{A}} & \text{and} & \hat{\underline{\mathbf{c}}} & \text{on} & \underline{\mathbf{M}},
\end{array}
$$

where $\underline{\mathbf{A}} = \underline{\mathbf{a}} \wedge \hat{\underline{\mathbf{a}}}$ etc.

Obviously we can therefore make any two lines on a conic its generating lines. It also follows that the lines $\underline{\mathbf{L}}$ and $\underline{\mathbf{M}}$ themselves are on the conic.

As a consequence, a line conic is defined by five distinct lines; dually, a point conic is defined by five distinct points.

At this point, we do not know much about the shape of a conic — all we know is that a conic is defined by a set of lines or, dually, by a set of points. But we can already make one claim: a nonsingular point conic may not be intersected by any line in more than two points.[2] For if such a line $\underline{\mathbf{K}}$ existed, it would have three pairs of pre-image/image lines intersecting on it. But then the projectivity between the two pencils must be a perspectivity, and the conic would be singular.

---

[2] A point conic is intersected by a line in a point $\underline{\mathbf{p}}$ if $\underline{\mathbf{p}}$ is both on the conic and on the line.

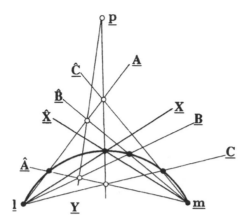

Figure 3.5. Point conic construction: a point on the conic is found by using the dual of Pappus' theorem.

3.2　Conic Constructions

It is now time to become more concrete: we know what it takes for a line to *be* on a conic, but how can we *find* one? Recall that a projectivity between two lines is defined by three image and pre-image point pairs. Pick three points $\underline{a}, \underline{b}, \underline{c}$ on $\underline{L}$ and $\hat{\underline{a}}, \hat{\underline{b}}, \hat{\underline{c}}$ on $\underline{M}$, such that their connecting lines are on Γ. How can we find more lines on Γ?

If $\underline{x}$ is an arbitrary point on $\underline{L}$, we have found a line on the conic as soon as we have found its image $\hat{\underline{x}}$ on $\underline{M}$ under the projectivity Φ that is defined by our three point pairs. But we already discussed how to find the image of an arbitrary point under a projectivity: this was done in Section 2.3 by means of Pappus' theorem. We would proceed exactly as in Fig. 2.4.

We may state this construction slightly differently: given five lines on a conic, find a sixth one. The given lines are $\underline{M}, \underline{L}, \underline{a} \wedge \hat{\underline{a}}, \underline{b} \wedge \hat{\underline{b}}$, and $\underline{c} \wedge \hat{\underline{c}}$.

The dual definition of conics as point conics leads to the dual construction; it is shown in Fig. 3.5. There, three concurrent lines $\underline{A}, \underline{B}, \underline{C}$ through $\underline{l}$ and their projective images $\hat{\underline{A}}, \hat{\underline{B}}, \hat{\underline{C}}$ through $\underline{m}$ are given. For an arbitrary line $\underline{X}$ through $\underline{l}$, we want to find its image $\hat{\underline{X}}$, so that we can find $\underline{x} = \underline{X} \wedge \hat{\underline{X}}$ on the conic. The dual of Pappus' theorem provides a construction: First, we construct the Pappus point $\underline{p}$ (see Section 1.9). Then we set

$$\underline{Y} = [\hat{\underline{A}} \wedge \underline{X}] \wedge \underline{p},$$

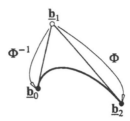

Figure 3.6. Points on a lince conic: $\underline{\mathbf{b}}_0$ and $\underline{\mathbf{b}}_2$ lie on Γ.

and find $\underline{\hat{\mathbf{X}}}$ as

$$\underline{\hat{\mathbf{X}}} = [\underline{\mathbf{A}} \wedge \underline{\mathbf{Y}}] \wedge \underline{\mathbf{m}}.$$

This construction may be viewed as a method to construct a sixth point on a conic once five points on it are given. This is a classical construction, and has been used by generations of draftsmen, see [5], [103], [102].

3.3   Points and Lines

What does it take for a point to be on a line conic? This very concept is not even defined yet, so here is a definition: A point $\underline{\mathbf{x}}$ is on a line conic Γ if there is exactly one line $\underline{\mathbf{L}}$ on Γ that contains $\underline{\mathbf{x}}$.

Dually, we may define a line on a point conic; such lines are the conic's *tangents*. A line $\underline{\mathbf{L}}$ is on a point conic Γ (i.e., $\underline{\mathbf{L}}$ is tangent to Γ), if there is exactly one point on Γ that lies on $\underline{\mathbf{L}}$.[3]

If a (line) conic Γ is defined by a projectivity Φ between two generating lines $\underline{\mathbf{L}}$ and $\underline{\mathbf{M}}$, then we know that those two lines are themselves on Γ.[4] We now ask: where does the conic touch $\underline{\mathbf{L}}$ and $\underline{\mathbf{M}}$, i.e., which are the points on the generating lines that are also on the conic? Let $\underline{\mathbf{b}}_1$ be the intersection of $\underline{\mathbf{L}}$ and $\underline{\mathbf{M}}$, as shown in Fig. 3.6. Since $\underline{\mathbf{b}}_1$ is on $\underline{\mathbf{L}}$, its image $\Phi\underline{\mathbf{b}}_1$ is on $\underline{\mathbf{M}}$. We call this image point $\underline{\mathbf{b}}_2$. Conversely, $\underline{\mathbf{b}}_1$ is also on $\underline{\mathbf{M}}$, and so its image $\Phi^{-1}\underline{\mathbf{b}}_1$ is on $\underline{\mathbf{L}}$. We define $\underline{\mathbf{b}}_0 = \Phi^{-1}\underline{\mathbf{b}}_1$.

We now claim that $\underline{\mathbf{b}}_0$ and $\underline{\mathbf{b}}_2$ are the two desired points of contact. Concentrating on $\underline{\mathbf{b}}_0$, we must show that a) $\underline{\mathbf{b}}_0$ is on Γ, and b) there is no other line on Γ that contains $\underline{\mathbf{b}}_0$. Since $\underline{\mathbf{b}}_0$ is the image of $\underline{\mathbf{b}}_1$ under Φ^{-1}, a) is satisfied. If another line on Γ existed that contains $\underline{\mathbf{b}}_0$, a second point — other than $\underline{\mathbf{b}}_1$ — must exist such that its image under Φ^{-1} is $\underline{\mathbf{b}}_0$. This is impossible, since Φ^{-1} is a bijection, and so our claim is verified.

---

[3]Note how we avoided any calculus here!

[4]See, for example, Fig. 3.4.

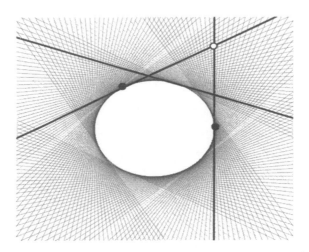

Figure 3.7. Conic constraints: a conic is given by two points, their corresponding tangents, and a third tangent.

Since we can generate a line conic from any two projectively related ranges, we are now able to find the *point of contact* on any line on the conic. For example, if we were given five lines to a conic, we are now able to find where they touch the conic. With five points on the conic, we may treat it as a point conic: we have thus shown that *every line conic can be expressed as a point conic and vice versa.* The "vice versa" part in this equivalence statement addresses the dual situation, of course. With this knowledge, we may refer to "lines on a conic" as "tangents to a conic".

In particular, we can state that a conic is defined by

a) two tangents $\underline{\mathbf{T}}_0, \underline{\mathbf{T}}_1$,

b) their points of contact $\underline{\mathbf{b}}_0, \underline{\mathbf{b}}_2$,

c) another tangent, defined by two points $\underline{\mathbf{q}}_0$ and $\underline{\mathbf{q}}_1$ on $\underline{\mathbf{T}}_0$ and $\underline{\mathbf{T}}_1$, respectively.

This tangent from c) is called the *shoulder tangent.* With $\underline{\mathbf{b}}_1$ being the intersection of $\underline{\mathbf{T}}_0$ and $\underline{\mathbf{T}}_1$, all we need to do is establish a projectivity between $\underline{\mathbf{T}}_0$ and $\underline{\mathbf{T}}_1$. It is given by the following pre-image and image point pairs:

$$\underline{\mathbf{b}}_0 \to \underline{\mathbf{b}}_1, \quad \underline{\mathbf{q}}_0 \to \underline{\mathbf{q}}_1, \quad \underline{\mathbf{b}}_1 \to \underline{\mathbf{b}}_2.$$

See Fig. 3.7 for an illustration and Chapter 4 for further elaboration.

In the context of point conics, we might have replaced the tangent $\underline{\mathbf{q}}_0 \wedge \underline{\mathbf{q}}_1$ by a *shoulder point,* $\underline{\mathbf{q}}$.

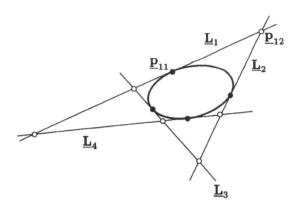

Figure 3.8. The four tangent theorem: four points are shown on each of four tangents to a conic. Their cross ratios are equal.

We are also in the position to formulate and prove another central theorem: the *four tangent theorem*. Consult Fig. 3.8 for the notation used. The theorem goes back to the projective development of conics by J. Steiner.

Let $\underline{\mathbf{L}}_i; i = 1, 2, 3, 4$, be four tangents to a conic Γ, and let $\underline{\mathbf{p}}_{i,i}$ their points of contact with Γ. Set

$$\underline{\mathbf{p}}_{i,j} = \underline{\mathbf{L}}_i \wedge \underline{\mathbf{L}}_j \quad \text{for} \quad i \neq j.$$

Then the following four cross ratios are all equal to a common constant c:

$$\text{cr}\big(\underline{\mathbf{p}}_{1,j}, \underline{\mathbf{p}}_{2,j}, \underline{\mathbf{p}}_{3,j}, \underline{\mathbf{p}}_{4,j}\big) = c; \quad j = 1, 2, 3, 4. \tag{3.2}$$

For a proof, we let $\underline{\mathbf{L}}_1$ and $\underline{\mathbf{L}}_4$ be generating lines of the conic. Then the four points on each of them are related by a projectivity. Since any pair of distinct lines $\underline{\mathbf{L}}_i$ and $\underline{\mathbf{L}}_j$ can also be used as generating lines, the theorem is proved.

The dual of the four tangent theorem will be called the *four point theorem* and can be stated like this: Let $\underline{\mathbf{p}}_i; i = 1, 2, 3, 4$, be four points on a (point) conic Γ, and let $\underline{\mathbf{L}}_{i,i}$ be the tangents to Γ at $\underline{\mathbf{p}}_i$. Set

$$\underline{\mathbf{L}}_{i,j} = \underline{\mathbf{p}}_i \wedge \underline{\mathbf{p}}_j \quad \text{for} \quad i \neq j.$$

Figure 3.9. The four point theorem: four points are shown as well as the four pencils generated by them. All four pencils are projectively related.

Then the following cross ratios are all equal to a common constant c:

$$\text{cr}(\underline{\mathbf{L}}_{1,j}, \underline{\mathbf{L}}_{2,j}, \underline{\mathbf{L}}_{3,j}, \underline{\mathbf{L}}_{4,j}) = c; \quad j = 1, 2, 3, 4. \tag{3.3}$$

This simply says that all four pencils through the $\underline{\mathbf{p}}_i$ are projectively related. For an illustration, see Fig. 3.9.

In that figure, let us concentrate on two of the points and the pencils through them, as shown in Fig. 3.10. If one pencil intersects a line $\underline{\mathbf{L}}$ and the other intersects a line $\underline{\mathbf{M}}$, both lines are related by a projectivity: if $\underline{\mathbf{x}}$ is a point on $\underline{\mathbf{L}}$, its image $\hat{\underline{\mathbf{x}}}$ is on $\underline{\mathbf{M}}$. We find $\hat{\underline{\mathbf{x}}}$ as follows: connect $\underline{\mathbf{x}}$ with $\underline{\mathbf{q}}$ and record the intersection $\underline{\mathbf{y}}$ of this line with Γ. Join $\underline{\mathbf{y}}$ with $\underline{\mathbf{p}}$, and $\hat{\underline{\mathbf{x}}}$ is the intersection of $\underline{\mathbf{p}} \wedge \underline{\mathbf{y}}$ with $\underline{\mathbf{M}}$.

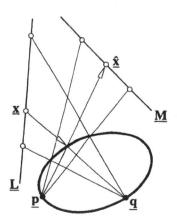

Figure 3.10. Conics and projectivities: the lines $\underline{\mathbf{L}}$ and $\underline{\mathbf{M}}$ are related by a conic-induced projectivity.

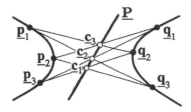

Figure 3.11. Pascal's theorem: the three hollow points are collinear.

3.4 Pascal's Theorem

Pascal's theorem was the first significant (*very* significant) advance that the theory of conics saw since it became stagnant in the Middle Ages. B. Pascal, who never received any formal training in mathematics, discovered it at the age of sixteen, in 1641. Pascal's theorem may be stated as follows:

Let $\underline{P}_1, \underline{P}_2, \underline{P}_3$ and $\underline{q}_1, \underline{q}_2, \underline{q}_3$ be any six points on a conic, as shown in Fig. 3.11. Then the three points

$$
\begin{aligned}
\underline{c}_1 &= (\underline{P}_2 \wedge \underline{q}_3) \wedge (\underline{P}_3 \wedge \underline{q}_2), \\
\underline{c}_2 &= (\underline{P}_1 \wedge \underline{q}_3) \wedge (\underline{P}_3 \wedge \underline{q}_1), \\
\underline{c}_3 &= (\underline{P}_1 \wedge \underline{q}_2) \wedge (\underline{P}_2 \wedge \underline{q}_1)
\end{aligned}
$$

are collinear. The line that they are on is called *Pascal line* $\underline{P}$.

For a proof, we turn to Fig. 3.12. First, we observe that we can relate the pencils through $\underline{q}_2$ and $\underline{q}_3$ by a projectivity — see the four point theorem above. This in turn generates a projectivity Φ between the lines $\underline{M}$ and $\underline{L}$. Since $\Phi \underline{q}_1 = \underline{q}_1$, we know that Φ is in fact a perspectivity, see Section 2.1.

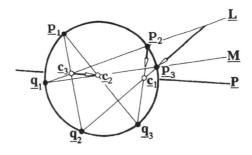

Figure 3.12. Pascal's theorem: the proof proceeds by showing that $\underline{L}$ and $\underline{M}$ are related by a perspectivity.

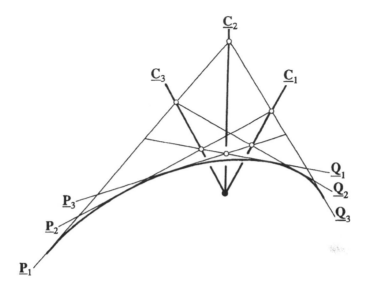

Figure 3.13. Brianchon's theorem: the three solid lines are concurrent.

Now let us determine Φc_3. To do this, we connect $\underline{c}_3$ with $\underline{q}_2$ and intersect the connecting line with Γ, yielding $\underline{p}_1$. Then we join $\underline{p}_1$ with $\underline{q}_3$ and intersect with $\underline{M}$: the image of $\underline{c}_3$ is $\underline{c}_2$. Thus the center of Φ must lie on $\underline{c}_1 \wedge \underline{c}_2$. We also know that $\Phi \underline{p}_2 \wedge \underline{q}_3$ must pass through Φ's center, which is therefore $\underline{c}_1$. Hence the three $\underline{c}_i$ are collinear. Since $\underline{c}_1$ is also on the line $\underline{q}_2 \wedge \underline{p}_3$, Pascal's theorem is proved.

Pascal's original proof proceeded in a different manner: he proved his theorem for a circle. Any other conic is a projective image of the circle, and the involved constructions are projectively invariant. Thus a proof for the circle is sufficient to prove the general case!

If the six points on the conic are not aligned as in our figure, Pascal's theorem still holds: as we relabel the points, there will be different Pascal lines.

If we dualize Pascal's theorem (see Section 1.9), we obtain *Brianchon's theorem:*

Let $\underline{P}_1, \underline{P}_2, \underline{P}_3$ and $\underline{Q}_1, \underline{Q}_2, \underline{Q}_3$ be any six tangents on a conic, as shown in Fig. 3.13. Then the three lines

$$
\begin{aligned}
\underline{C}_1 &= (\underline{P}_2 \wedge \underline{Q}_3) \wedge (\underline{P}_3 \wedge \underline{Q}_2), \\
\underline{C}_2 &= (\underline{P}_1 \wedge \underline{Q}_3) \wedge (\underline{P}_3 \wedge \underline{Q}_1), \\
\underline{C}_3 &= (\underline{P}_1 \wedge \underline{Q}_2) \wedge (\underline{P}_2 \wedge \underline{Q}_1)
\end{aligned}
$$

are concurrent. The point that they meet in is called *Brianchon point.*

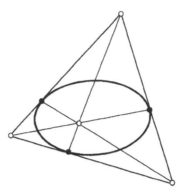

Figure 3.14. Brianchon's theorem: in this special case, a triangle is circumscribed around a conic. The three "medians" are concurrent.

We do not have to prove this theorem due to the principle of duality. When Brianchon proved his theorem — in 1806 — this principle had not yet been discovered.

A special case of Brianchon's theorem is obtained when we let the six points approach each other, resulting in three points together with their tangents. We then have: if a triangle is circumscribed around a conic, the lines through points of contact and opposite triangle vertices are concurrent.[5] See Fig. 3.14 for an illustration.

Pascal's theorem makes an assertion about six points known to be on a conic. If we retrace the proof of Pascal's theorem "backwards", we see that the following is also true: if six points $\underline{\mathbf{p}}_i, \underline{\mathbf{q}}_i$, $i = 1, 2, 3$, determine three collinear points $\underline{\mathbf{c}}_i = \underline{\mathbf{p}}_j \wedge \underline{\mathbf{p}}_k$, with i, j, k all different, then the six points lie on a conic.

Finally, a special case of Pascal's theorem: referring to Fig. 3.15, we let two point pairs coincide: we set $\underline{\mathbf{p}}_1 = \underline{\mathbf{q}}_3$ and $\underline{\mathbf{p}}_3 = \underline{\mathbf{q}}_1$. Then the lines $\underline{\mathbf{p}}_1 \wedge \underline{\mathbf{q}}_3$ and $\underline{\mathbf{p}}_3 \wedge \underline{\mathbf{q}}_1$ become tangents. The Pascal line now goes through the Pappus point $\underline{\mathbf{p}}$ of the two pencils through $\underline{\mathbf{p}}_1 = \underline{\mathbf{q}}_3$ and $\underline{\mathbf{p}}_3 = \underline{\mathbf{q}}_1$.[6]

This leads to one of the classical conic constructions: given three points and two tangents to a conic, find more points on it. Again referring to Fig. 3.15, assume the tangents $\underline{\mathbf{T}}_0$ and $\underline{\mathbf{T}}_1$ in $\underline{\mathbf{p}}_1 = \underline{\mathbf{q}}_3$ and $\underline{\mathbf{p}}_3 = \underline{\mathbf{q}}_1$ are given, plus the point $\underline{\mathbf{p}}_2$. We can now fix an arbitrary Pascal line $\underline{\mathbf{L}}$ and

---

[5] A simple proof for this special case is as follows: the statement holds for an equilateral triangle circumscribed around a circle. All other cases follow by projectively mapping this configuration.

[6] The Pappus point would normally depend on the ordering of the points. In this special situation, all orderings still give the same point.

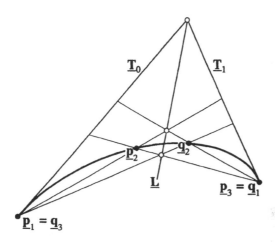

Figure 3.15. Pascal's theorem: a special case.

then construct the point $\underline{q}_2$ defined by it. As we let the Pascal line vary, we generate more points on the conic. Since we can write

$$\underline{L} = (1 - t)\underline{T}_0 + t\underline{T}_1,$$

we have found a *parametrization* of our conic: for every value of t, we can find a point $\underline{q}_2(t)$ on the conic. In the next chapter, we will see a more elegant parametric form for conics.

3.5 Affine Conics

When we consider the extended affine plane, we know that affine maps leave the line $\mathbf{L}_\infty$ invariant, according to the development in Section 2.7. A conic may have zero, one, or two intersections with $\mathbf{L}_\infty$, and this number would be unchanged by an affine map. This leads to the following classification of conics in the extended affine plane: *a conic is called an ellipse, a parabola, or a hyperbola, if it has zero, one, or two intersections with $\mathbf{L}_\infty$, respectively.* Since $\mathbf{L}_\infty$ is the image of some line $\underline{H}$ under a projective map from $I\!\!P^2$, we arrive at the illustration 3.16.

Before we look at each affine conic type, a definition: a curve is called *convex* if, for any tangent to it, the whole curve lies on one side of this tangent. Note that this definition only makes sense in an affine context! See Section 1.11 for an explanation.

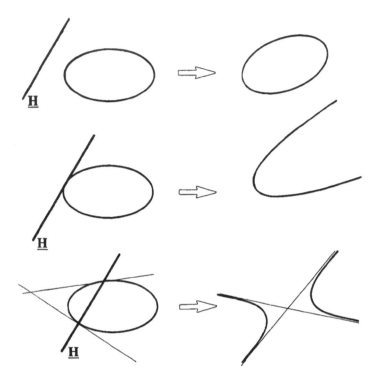

Figure 3.16. Affine conics: from top to bottom: ellipse, parabola, hyperbola. In each case, **H** is the pre-image of the line at infinity.

The **ellipse**, having no points in common with the line at infinity, is a finite curve. Since it cannot be intersected by any line more than twice, it is a *convex* curve.

Four points are sufficient to determine an ellipse. The points must be vertices of a convex quadrilateral. In order to produce a numerically reliable ellipse, the four points should not be "too close" together.

The **parabola** has exactly one point in common with L_∞, hence L_∞ is a tangent of the parabola. It is also a convex curve, although not a finite one.

Since we know one tangent of the parabola, four more points p_i will determine it completely. But these points cannot be arbitrarily located since a parabola is a convex curve. The points must be vertices of a convex quadrilateral — see also [99].

The **hyperbola** intersects L_∞ twice. Let $\underline{x}$ and $\underline{y}$ be the intersections with the hyperbolas' pre-image with $\underline{L}$. The tangents at $\underline{x}$ and $\underline{y}$ are mapped to the *asymptotes* of the hyperbola. A hyperbola is not a convex curve.

A hyperbola is determined by its two asymptotes and one more point.

3.6 Problems

1. Show that two conics cannot intersect in more than four points.

2. Specialize the situation from Fig. 3.15 even more, namely by considering three points on a conic and their respective tangents.

3. Develop the construction of a point conic using a given shoulder point.

4. What are the models of line and point conics in the spherical or circular models of the projective plane? If you have access to the appropriate kind of graphics equipment, illustrate!

4

Conics in Parametric Form

We have so far covered the basic theoretical aspects of conics in the projective plane. Our definition was based on the concept of projectivities between lines or pencils — we will now use that concept for a more tangible description. This will be the parametric form, which will allow us to trace curves, and also to calculate derivatives. We will spend some time on parametric conics, as they are important by themselves, but also since many of their properties will carry over very naturally to rational Bézier curves or to NURBS.

4.1 Parametric Curves in $I\!\!P^2$

Before we start to treat conics in parametric form, we present a few facts about parametric curves in projective space; some phenomena arise here that are quite different from the more familiar affine treatment of curves.

A parametric curve $\underline{x}(t)$ is a map from the projective line — endowed with a parameter t — into $I\!\!P^2$. Let us consider the derivative of $\underline{x}(t)$. We assume that each of the components of $\underline{x}$ is differentiable with respect to t. Let Δt be a nonzero number. Then the three collinear points $\underline{x}(t), \underline{x}(t + \Delta t), \underline{x}(t + \Delta t) - \underline{x}(t)$ are on a *secant* of the curve, as shown in Fig. 4.1.

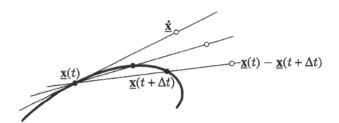

Figure 4.1. Projective derivatives: as $\Delta t \to 0$, the secants approach the tangent at $\underline{\mathbf{x}}(t)$.

Note that

$$\underline{\mathbf{x}}(t + \Delta t) - \underline{\mathbf{x}}(t) \; \hat{=} \; \frac{\underline{\mathbf{x}}(t + \Delta t) - \underline{\mathbf{x}}(t)}{\Delta t}.$$

We now take the limit $\Delta t \to 0$:

$$\underline{\dot{\mathbf{x}}} := \frac{\mathrm{d}\underline{\mathbf{x}}(t)}{\mathrm{d}t} = \lim_{\Delta t \to 0} \frac{\underline{\mathbf{x}}(t + \Delta t) - \underline{\mathbf{x}}(t)}{\Delta t}.$$

In this limit process, the secant approaches the tangent $\underline{\mathbf{T}}(t)$ and the point $\big(\underline{\mathbf{x}}(t + \Delta t) - \underline{\mathbf{x}}(t)\big)/\Delta t$ approaches the *derivative* $\underline{\dot{\mathbf{x}}}(t)$.

We are used to thinking of derivatives as tangent *vectors*, but in the projective plane, there are no vectors, and hence the derivative is yet another point. This point is located on the tangent $\underline{\mathbf{T}}(t)$, which is thus given by

$$\underline{\mathbf{T}} = \underline{\mathbf{x}} \wedge \underline{\dot{\mathbf{x}}}. \tag{4.1}$$

Note how this results coincides with the derivative (1.17) of a parametric straight line.

Similarly, the second derivative $\underline{\ddot{\mathbf{x}}}$ is yet another point in $I\!\!P^2$. If, for some t, the three points $\underline{\mathbf{x}}, \underline{\dot{\mathbf{x}}}, \underline{\ddot{\mathbf{x}}}$ are collinear, i.e., if $\det[\underline{\mathbf{x}}, \underline{\dot{\mathbf{x}}}, \underline{\ddot{\mathbf{x}}}] = 0$, then the curve is said to have an *inflection point* at $\underline{\mathbf{x}}(t)$.[1] In projective differential geometry, one treats inflection points as special cases — then every (non-inflection) point on the curve and every higher derivative may be written as a linear combination of $\underline{\mathbf{x}}, \underline{\dot{\mathbf{x}}}, \underline{\ddot{\mathbf{x}}}$, in fashion similar to the concept of a Frenet frame in classical differential geometry. See [27], [160].

Another interesting phenomenon arises from the observation that setting $\underline{\mathbf{y}}(t) = \rho(t)\underline{\mathbf{x}}(t)$ results in $\underline{\mathbf{y}}(t) \hat{=} \underline{\mathbf{x}}(t)$ for all t. Derivatives change, however, due to application of the product rule. For the first two derivatives,

---

[1]In this chapter, we shall only consider conics; these do not have inflection points.

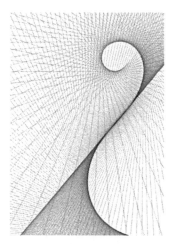

Figure 4.2. Line curves: instead of viewing a curve as the locus of a moving point, it may be viewed as the locus of a moving line.

we obtain (following Geise and Nestler[2]):

$$[\underline{\mathbf{y}}, \underline{\dot{\mathbf{y}}}, \underline{\ddot{\mathbf{y}}}] = [\mathbf{x}, \dot{\mathbf{x}}, \ddot{\mathbf{x}}] \begin{bmatrix} \rho & \dot{\rho} & \ddot{\rho} \\ 0 & \rho & 2\dot{\rho} \\ 0 & 0 & \rho \end{bmatrix}. \tag{4.2}$$

Similarly, let $t = \phi(s)$ represent a parameter transformation (this requires $\dot{\phi}(s) > 0$ for all s). Then we set $\underline{\mathbf{z}}(s) = \mathbf{x}(\phi(s)) = \mathbf{x}(t)$ and obtain, using the chain and product rules:

$$[\underline{\mathbf{z}}, \underline{\dot{\mathbf{z}}}, \underline{\ddot{\mathbf{z}}}] = [\mathbf{x}, \dot{\mathbf{x}}, \ddot{\mathbf{x}}] \begin{bmatrix} 1 & 0 & 0 \\ 0 & \phi' & \phi'' \\ 0 & 0 & \phi'^2 \end{bmatrix}, \tag{4.3}$$

where the prime $'$ denotes derivatives with respect to s.

We finish with the observation that curves may not only be the locus of points. Dually, they may also be the locus of *lines* — meaning they may be viewed as consisting of a set of lines which touch the curve. These lines are the curve's tangents. For an illustration, see Fig. 4.2.

---

[2]Private communication.

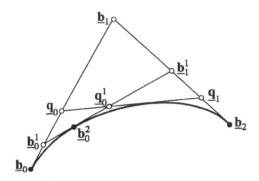

Figure 4.3. The Bézier form of conics: the conic is defined by the control polygon $\underline{\mathbf{b}}_0, \underline{\mathbf{b}}_1, \underline{\mathbf{b}}_2$ and a shoulder tangent.

4.2   The Bernstein Form of a Conic

Let $\underline{\mathbf{b}}_0$ and $\underline{\mathbf{b}}_2$ be two distinct points on a conic Γ. Let the corresponding tangents $\underline{\mathbf{B}}_0$ and $\underline{\mathbf{B}}_2$ intersect in the point $\underline{\mathbf{b}}_1$. Let

$$\underline{\mathbf{q}}_0 = \frac{1}{2}\underline{\mathbf{b}}_0 + \frac{1}{2}\underline{\mathbf{b}}_1 \quad \text{and} \quad \underline{\mathbf{q}}_1 = \frac{1}{2}\underline{\mathbf{b}}_1 + \frac{1}{2}\underline{\mathbf{b}}_2$$

be two points on these tangents, such that $\underline{\mathbf{Q}} = \underline{\mathbf{q}}_0 \wedge \underline{\mathbf{q}}_1$ is another tangent to Γ, see Fig. 4.3. [3] We call the tangent $\underline{\mathbf{Q}}$ the *shoulder tangent*.

We know that this information — two points plus three tangents — is sufficient to define the conic. Let

$$\underline{\mathbf{b}}_0^1 = \underline{\mathbf{b}}_0^1(t) = (1-t)\underline{\mathbf{b}}_0 + t\underline{\mathbf{b}}_1 \tag{4.4}$$

a point on $\underline{\mathbf{B}}_0$. We will now locate a point $\underline{\mathbf{b}}_1^1$ on $\underline{\mathbf{B}}_2$ such that $\underline{\mathbf{T}} = \underline{\mathbf{b}}_0^1 \wedge \underline{\mathbf{b}}_1^1$ is tangent to Γ. From the four tangent theorem (3.2) we get the following cross ratio identity:

$$\mathrm{cr}(\underline{\mathbf{b}}_0, \underline{\mathbf{b}}_0^1, \underline{\mathbf{q}}_0, \underline{\mathbf{b}}_1) = \mathrm{cr}(\underline{\mathbf{b}}_1, \underline{\mathbf{b}}_1^1, \underline{\mathbf{q}}_1, \underline{\mathbf{b}}_2).$$

Since

$$\mathrm{cr}(\underline{\mathbf{b}}_0, \underline{\mathbf{b}}_0^1, \underline{\mathbf{q}}_0, \underline{\mathbf{b}}_1) = \frac{t}{1-t},$$

It follows that $\underline{\mathbf{b}}_1^1$ is given by

$$\underline{\mathbf{b}}_1^1 = \underline{\mathbf{b}}_1^1(t) = (1-t)\underline{\mathbf{b}}_1 + t\underline{\mathbf{b}}_2. \tag{4.5}$$

---

[3] Keep in mind that this does not restrict the location of the $\underline{\mathbf{q}}_i$, as demonstrated in Example 1.1!

If we are interested in the tangent $\mathbf{T}$ to Γ, we are finished (for an equation of the tangent, see (4.10). But we can go one step further and ask: what is the point of contact — we will call it $\underline{\mathbf{b}}_0^2(t)$ — between $\mathbf{T}$ and Γ? Let us define

$$\underline{\mathbf{q}}_0^1 = \underline{\mathbf{q}}_0^1(t) = \underline{\mathbf{T}}(t) \wedge \underline{\mathbf{Q}}.$$

Since $\underline{\mathbf{q}}_0^1$ must satisfy both

$$\underline{\mathbf{q}}_0^1 = (1 - \alpha)\underline{\mathbf{b}}_0^1 + \alpha\underline{\mathbf{b}}_1^1 \quad \text{and} \quad \underline{\mathbf{q}}_0^1 = (1 - \beta)\underline{\mathbf{q}}_0 + \beta\underline{\mathbf{q}}_1,$$

we find that

$$\underline{\mathbf{q}}_0^1 = \frac{1}{2}\underline{\mathbf{b}}_0^1 + \frac{1}{2}\underline{\mathbf{b}}_1^1. \tag{4.6}$$

Invoking the four tangent theorem again, we see that we must have

$$\mathrm{cr}(\underline{\mathbf{b}}_0^1, \underline{\mathbf{b}}_0^2, \underline{\mathbf{q}}_0^1, \underline{\mathbf{b}}_1^1) = \frac{t}{1 - t},$$

yielding

$$\underline{\mathbf{b}}_0^2(t) = (1 - t)\underline{\mathbf{b}}_0^1(t) + t\underline{\mathbf{b}}_1^1(t). \tag{4.7}$$

We can now insert (4.4) and (4.5) into (4.7):

$$\underline{\mathbf{b}}_0^2(t) = (1 - t)^2\underline{\mathbf{b}}_0 + 2t(1 - t)\underline{\mathbf{b}}_1 + t^2\underline{\mathbf{b}}_2. \tag{4.8}$$

Defining the quadratic *Bernstein polynomials* B_i^2 as

$$B_i^2(t) = \binom{2}{i}t^i(1 - t)^{2-i}; \quad i = 0, 1, 2,$$

we can write our conic as a *projective quadratic Bézier curve:*

$$\underline{\mathbf{b}}_0^2(t) = \sum_{i=0}^{2} \underline{\mathbf{b}}_i B_i^2(t). \tag{4.9}$$

This form of the conic is a *parametric quadratic curve*. As the parameter t traces out all values between $+\infty$ and $-\infty$, the point $\underline{\mathbf{b}}_0^2(t)$ traces out all of the conic. We also use the term "quadratic Bézier curve". This terminology will be explained in Chapter 7.

Our recursive algorithm, called the *de Casteljau algorithm*, is conveniently written in a triangular arrangement as shown in Example 4.1.

We should note that, although we worked in the context of line conics — using the four-tangent theorem — we finally arrived at the equation of

$$
\begin{array}{lll}
\underline{\mathbf{b}}_0 & & (1-t) \quad \bullet \\
\underline{\mathbf{b}}_1 & \underline{\mathbf{b}}_0^1 & \qquad\qquad\qquad \searrow \\
\underline{\mathbf{b}}_2 & \underline{\mathbf{b}}_1^1 & \underline{\mathbf{b}}_0^2 \qquad\quad t \quad\; \bullet \;\rightarrow\; + \;\rightarrow\; \bullet
\end{array}
$$

$$
\begin{bmatrix} 0 \\ 0 \\ 1 \end{bmatrix}
$$

$$
\begin{bmatrix} 0 \\ 2 \\ 1 \end{bmatrix} \quad \begin{bmatrix} 0 \\ 1 \\ 1 \end{bmatrix}
$$

$$
\begin{bmatrix} 8 \\ 2 \\ 5 \end{bmatrix} \quad \begin{bmatrix} 4 \\ 2 \\ 3 \end{bmatrix} \quad \begin{bmatrix} 2 \\ \frac{3}{2} \\ 2 \end{bmatrix}
$$

Example 4.1. The triangular scheme for the conic de Casteljau algorithm. Top left: the overall scheme; top right: the rule to generate it. Bottom: a numerical example for $t = \frac{1}{2}$.

a *point* on the conic. For the corresponding tangent, $\underline{\mathbf{T}}$, we write

$$
\begin{aligned}
\underline{\mathbf{T}}(t) &= \underline{\mathbf{b}}_0^1 \wedge \underline{\mathbf{b}}_1^1 \\
&= [(1-t)\underline{\mathbf{b}}_0 + t\underline{\mathbf{b}}_1] \wedge [(1-t)\underline{\mathbf{b}}_1 + t\underline{\mathbf{b}}_2] \\
&= (1-t)^2[\underline{\mathbf{b}}_0 \wedge \underline{\mathbf{b}}_1] + t(1-t)[\underline{\mathbf{b}}_0 \wedge \underline{\mathbf{b}}_2] + \\
&\quad\; t(1-t)[\underline{\mathbf{b}}_1 \wedge \underline{\mathbf{b}}_1] + t^2[\underline{\mathbf{b}}_1 \wedge \underline{\mathbf{b}}_2].
\end{aligned}
$$

We observe that $\underline{\mathbf{b}}_1 \wedge \underline{\mathbf{b}}_1 = \mathbf{0}$ and arrive at

$$
\underline{\mathbf{T}}(t) = (1-t)^2\underline{\mathbf{B}}_0 + t(1-t)\underline{\mathbf{B}}_1 + t^2\underline{\mathbf{B}}_2 \tag{4.10}
$$

for the equation of the tangent, expressed in terms of the control polygon legs

$$
\underline{\mathbf{B}}_0 = \underline{\mathbf{b}}_0 \wedge \underline{\mathbf{b}}_1, \quad \underline{\mathbf{B}}_1 = \underline{\mathbf{b}}_0 \wedge \underline{\mathbf{b}}_2, \quad \underline{\mathbf{B}}_2 = \underline{\mathbf{b}}_1 \wedge \underline{\mathbf{b}}_2. \tag{4.11}
$$

For the special case $t = \frac{1}{2}$, we obtain

$$
\underline{\mathbf{q}}_0 \wedge \underline{\mathbf{q}}_1 = \underline{\mathbf{B}}_0 + \underline{\mathbf{B}}_1 + \underline{\mathbf{B}}_2.
$$

4.3   The Dual Form

Our above development was in the context of line conics: we started with the line conic definition and arrived at the point-type definition (4.9). We

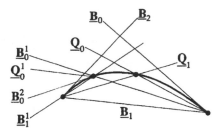

Figure 4.4. Dual conic definition: two pencils and an additional point are given.

should also be able to take the dual approach. We would then be given two points $\underline{b}_0$ and $\underline{b}_2$ and the respective tangents $\underline{B}_0$ and $\underline{B}_2$, see Fig. 4.4. In addition, another point $\underline{q}$ on the conic would be given, defining two lines $\underline{Q}_0 = \underline{q} \wedge \underline{b}_0$ and $\underline{Q}_1 = \underline{q} \wedge \underline{b}_2$.

We need three pairs of lines from the pencils through $\underline{b}_0$ and $\underline{b}_2$, thus defining a projectivity Φ between the two pencils. We select, with the notation from Fig. 4.4,

$$\underline{B}_2 = \Phi\underline{B}_1, \quad \underline{Q}_1 = \Phi\underline{Q}_0, \quad \underline{B}_1 = \Phi\underline{B}_0.$$

Again, we set $\underline{Q}_0 = \frac{1}{2}\underline{B}_0 + \frac{1}{2}\underline{B}_1$ and $\underline{Q}_1 = \frac{1}{2}\underline{B}_1 + \frac{1}{2}\underline{B}_2$. Selecting an arbitrary line

$$\underline{B}_0^1(t) = (1-t)\underline{B}_0 + t\underline{B}_1$$

in the pencil through $\underline{b}_0$, and utilizing the four point theorem (3.3), we find its image under Φ to be $\underline{B}_1^1(t) = (1-t)\underline{B}_1 + t\underline{B}_2$.

We may set $\underline{Q}_0^1 = \frac{1}{2}\underline{B}_0^1 + \frac{1}{2}\underline{B}_1^1$ in analogy to (4.6). We can now apply the four point theorem to the pencil of lines through $\underline{B}_0^1 \wedge \underline{B}_1^1$, and find that the tangent $\underline{B}_0^2(t)$ to the conic must satisfy $\underline{B}_0^2 = (1-t)\underline{B}_0^1 + t\underline{B}_1^1$, resulting in

$$\underline{B}_0^2(t) = \sum_{i=0}^{2} \underline{B}_i B_i^2(t), \tag{4.12}$$

which is of course the dual of (4.9) and equivalent to (4.10). This is the parametric *line conic* form; in this form, we refer to the conic as a *control line conic*. We will encounter this type of equation again in Section 7.12. Fig. 4.5 illustrates.

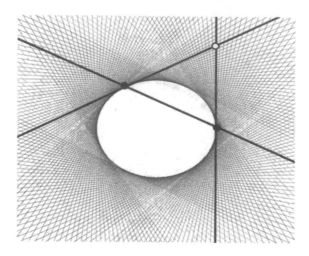

Figure 4.5. Control line conics: the conic is defined by three control lines (bold). They define the set of tangents that envelope the conic.

4.4 Derivatives

Now for the derivative of a projective conic in Bézier form: it is given by

$$\dot{\underline{\mathbf{b}}}(t) = 2(1 - t)\Delta\underline{\mathbf{b}}_0 + 2t\Delta\underline{\mathbf{b}}_1,$$

or, equivalently,

$$\dot{\underline{\mathbf{b}}}(t) \hat{=} (1 - t)\Delta\underline{\mathbf{b}}_0 + t\Delta\underline{\mathbf{b}}_1 \tag{4.13}$$

where Δ is the *forward difference operator:*

$$\Delta\underline{\mathbf{b}}_i = \underline{\mathbf{b}}_{i+1} - \underline{\mathbf{b}}_i.$$

Since both $\Delta\underline{\mathbf{b}}_0$ and $\Delta\underline{\mathbf{b}}_1$ are points, it follows that (4.13) describes a line, and we have that *all derivatives of a conic are collinear.* This is to be expected, of course, since the derivative of a quadratic is linear. Fig. 4.6 shows a conic and its first derivative curve.

Where is this line that all conic derivatives are on? Rewrite (4.9) as

$$\underline{\mathbf{b}}(t) = \underline{\mathbf{b}}_0 + 2t\Delta\underline{\mathbf{b}}_0 + t^2\Delta^2\underline{\mathbf{b}}_0, \tag{4.14}$$

with

$$\Delta^2\underline{\mathbf{b}}_0 = \Delta\underline{\mathbf{b}}_1 - \Delta\underline{\mathbf{b}}_0. \tag{4.15}$$

We now see that

$$\underline{\mathbf{b}}(\infty) := \lim_{t \to \infty} \underline{\mathbf{b}}(t) \hat{=} \Delta^2\underline{\mathbf{b}}_0. \tag{4.16}$$

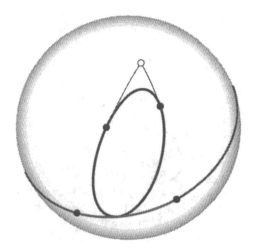

Figure 4.6. Conic derivatives: a conic (black) and its first derivative curve (gray).

Using the same limit argument again, we obtain

$$\underline{\mathbf{b}}(\infty)\hat{=}\dot{\underline{\mathbf{b}}}(\infty)\hat{=}\ddot{\underline{\mathbf{b}}}(\infty),\tag{4.17}$$

where $\ddot{\underline{\mathbf{b}}}$ stands for the second derivative. Note that the second derivative is a point on the conic!

Since $\Delta\underline{\mathbf{b}}_0, \Delta^2\underline{\mathbf{b}}_0$, and $\Delta\underline{\mathbf{b}}_1$ are collinear points by virtue of (4.15), it follows from (4.17) that all first derivatives to $\underline{\mathbf{b}}$ lie on the tangent corresponding to $t = \infty$:

$$\dot{\underline{\mathbf{b}}}(t) = \underline{\mathbf{T}}(\infty) \wedge \underline{\mathbf{T}}(t).\tag{4.18}$$

For an illustration, see Fig. 4.7.

That figure suggests another identity:

$$\det[\underline{\mathbf{b}}_1, \underline{\mathbf{b}}(\tfrac{1}{2}), \underline{\mathbf{b}}(\infty)] = 0,\tag{4.19}$$

i.e., the three points $\underline{\mathbf{b}}_1, \underline{\mathbf{b}}(\tfrac{1}{2}), \underline{\mathbf{b}}(\infty)$ are collinear. To prove this fact, we verify the simple identity

$$\underline{\mathbf{b}}(\infty) - 2\underline{\mathbf{b}}(\tfrac{1}{2}) = -\frac{5}{2}\underline{\mathbf{b}}_1.$$

The endpoints of a conic segment deserve special attention: their derivatives are given by

$$\dot{\underline{\mathbf{b}}}(0) = 2\Delta\underline{\mathbf{b}}_0 \quad \text{and} \quad \dot{\underline{\mathbf{b}}}(1) = 2\Delta\underline{\mathbf{b}}_1.\tag{4.20}$$

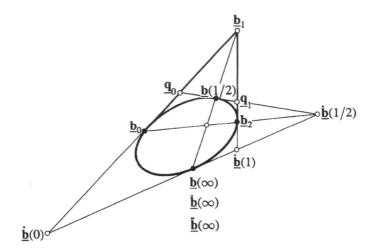

Figure 4.7. Tangents and derivatives: several interesting cases.

Note that the four points $\underline{\mathbf{b}}_0, \underline{\dot{\mathbf{q}}}_0, \underline{\mathbf{b}}_1, \underline{\dot{\mathbf{b}}}(0)$ are harmonic, as are the points $\underline{\mathbf{b}}_1, \underline{\dot{\mathbf{q}}}_1, \underline{\mathbf{b}}_2, \underline{\dot{\mathbf{b}}}(1)$. This follows directly from (2.6).

4.5   The Implicit Form

The parametric form of a conic offers a means to compute points on it: given a parameter value t, we can find the corresponding point on the conic. On the other hand, it offers no solution to the following problem: given a point $\underline{\mathbf{x}}$, is it on the conic or not? The implicit form will help answer this question. We shall first derive the implicit form for a special case of parametric conics, then we will take it to the general case.

Let us assign the following coordinates to the defining points of the conic Γ:

$$\underline{\mathbf{b}}_0 = \begin{bmatrix} 0 \\ 0 \\ 1 \end{bmatrix}, \ \underline{\mathbf{b}}_1 = \begin{bmatrix} 0 \\ 1 \\ 0 \end{bmatrix}, \ \underline{\mathbf{b}}_2 = \begin{bmatrix} 1 \\ 0 \\ 0 \end{bmatrix}, \ \underline{\mathbf{q}}_0 = \frac{1}{2}\begin{bmatrix} 0 \\ 1 \\ 1 \end{bmatrix}, \ \underline{\mathbf{q}}_1 = \frac{1}{2}\begin{bmatrix} 1 \\ 1 \\ 0 \end{bmatrix}.$$

Then Γ takes on the simple form

$$\underline{\mathbf{x}} = \underline{\mathbf{x}}(t) = \begin{bmatrix} t^2 \\ 2t(1-t) \\ (1-t)^2 \end{bmatrix}, \tag{4.21}$$

from which we deduce the identity

$$y^2 = 4xz. \tag{4.22}$$

This expression — the implicit form of a conic — does not involve the parameter value t any more! It holds for any point on the conic, regardless of its parameter value.

We can rewrite (4.22) in matrix form:

$$\underline{\mathbf{x}}^T A \underline{\mathbf{x}} = 0, \tag{4.23}$$

with the symmetric matrix

$$A = \begin{bmatrix} 0 & 0 & -2 \\ 0 & 1 & 0 \\ -2 & 0 & 0 \end{bmatrix}.$$

We can now decide if a given point is on Γ just by checking if its coordinates satisfy (4.23) or not.

Let us define the line

$$\underline{\mathbf{T}} = \underline{\mathbf{x}}^T A. \tag{4.24}$$

Then $\underline{\mathbf{T}}\underline{\mathbf{x}} = 0$ by (4.23), and so $\underline{\mathbf{x}}$ lies on $\underline{\mathbf{T}}$. In terms of the parameter t, we can express $\underline{\mathbf{T}}$ as

$$\underline{\mathbf{T}} = 2[-(1-t)^2, t(1-t), -t^2].$$

In our special coordinate system, we have

$$\underline{\mathbf{b}}_0^1 = \begin{bmatrix} 0 \\ t \\ 1-t \end{bmatrix} \quad \text{and} \quad \underline{\mathbf{b}}_1^1 = \begin{bmatrix} t \\ 1-t \\ 0 \end{bmatrix}.$$

It is now easy to check that $\underline{\mathbf{T}}\underline{\mathbf{b}}_0^1 = \underline{\mathbf{T}}\underline{\mathbf{b}}_1^1 = 0$, i.e., the three collinear points $\underline{\mathbf{b}}_0^1, \underline{\mathbf{b}}_0^2, \underline{\mathbf{b}}_1^1$ all are on $\underline{\mathbf{T}}$. Thus $\underline{\mathbf{T}}$ is the tangent to Γ at $\underline{\mathbf{x}}$!

So far, our implicit approach lacks generality: it seems to rely heavily on the special coordinate system that we have used. We will now show that this is not so. Let $\underline{\mathbf{a}}, \underline{\mathbf{b}}, \underline{\mathbf{c}}, \underline{\mathbf{d}}$ form a projective reference frame. With respect to it, the point $\underline{\mathbf{x}}$ has coordinates $\underline{\hat{\mathbf{x}}}$, given by a coordinate transformation $\underline{\hat{\mathbf{x}}} = M\underline{\mathbf{x}}$, where M is a nonsingular 3×3 matrix. Then the implicit form of our conic Γ becomes

$$\underline{\mathbf{x}}^T M^T A M \underline{\mathbf{x}} = 0. \tag{4.25}$$

Since $M^T A M$ is again a nonsingular symmetric 3×3 matrix, we see that no matter what coordinate system we use, a conic can always be written as $\underline{x}^T B \underline{x} = 0$, with B a nonsingular symmetric 3×3 matrix. If $\underline{x}$ is a point on this conic, then $\underline{x}^T A$ is the corresponding tangent.

We may use the implicit form to talk about points "inside" and "outside" the conic: define all points that satisfy $\underline{x}^T A \underline{x} < 0$ to be inside the conic, and all those with $\underline{x}^T A \underline{x} > 0$ to be outside.[4] This definition is invariant under projective maps and allows us to split up the projective plane into two separate parts.

Another way of specifying "inside" and "outside" is the following. We call a point $\underline{p}$ outside the conic if we can draw two tangents from $\underline{p}$ to the conic. The point is inside the conic if no tangents can be drawn from it. If we want to decide upon the position of a point, we form the *polar* $\underline{p}^T A$; see also Section 4.8. If this line has two intersections $\underline{x}_1$ and $\underline{x}_2$ with the conic, then the corresponding tangents will both go through $\underline{p}$. We thus need to solve the equation

$$\underline{x}_1 \wedge \underline{x}_2 = \underline{p}^T A$$

for $\underline{x}_1$ and $\underline{x}_2$.

In the Problems section the implicit form of an affine conic is given by (4.40); the implicit form of a projective line conic is given by (4.41).

We have shown in this section how to obtain the implicit form of a conic once we have a parametric representation of it. How about the other way around: can we go from the implicit form to the parametric form? We can, but not in a unique way. The implicit form does not single out any points on the conic, whereas the parametric one does: we have to specify two points, $\underline{b}_0$ and $\underline{b}_2$ on the conic, together with their tangents, plus a third tangent. So if we are given the implicit form, we first have to determine two points on it, and then find their tangents. This can be done using (4.24). The interplay between parametric and implicit forms, also for higher degree curves, has recently attracted some attention. We list: [1], [2],[3], [80], [81], [134], [138].

4.6   Implicit Interpolating Conics

We know that five points determine a conic. What is its implicit form? To answer that question, assume that the points are $\underline{x}_1, \ldots, \underline{x}_5$. Any point $\underline{x}$

---

[4] Of course, we could define this the other way around as well!

on the desired conic must satisfy

$$
\begin{vmatrix}
x^2 & y^2 & z^2 & xy & xz & yz \\
x_1^2 & y_1^2 & z_1^2 & x_1y_1 & x_1z_1 & y_1z_1 \\
x_2^2 & y_2^2 & z_2^2 & x_2y_2 & x_2z_2 & y_2z_2 \\
x_3^2 & y_3^2 & z_3^2 & x_3y_3 & x_3z_3 & y_3z_3 \\
x_4^2 & y_4^2 & z_4^2 & x_4y_4 & x_4z_4 & y_4z_4 \\
x_5^2 & y_5^2 & z_5^2 & x_5y_5 & x_5z_5 & y_5z_5
\end{vmatrix} = 0. \tag{4.26}
$$

Expand this determinant by the first row and you see that the above expression is quadratic in x, y, z; thus we can bring it into the form (4.25). Since it vanishes for $\underline{\mathbf{x}} = \underline{\mathbf{x}}_i; i = 1, \ldots, 5$, it interpolates to the given points.

Five *affine* points $\mathbf{x}_1, \ldots, \mathbf{x}_5$ also determine a conic; its implicit form is given by

$$
\begin{vmatrix}
x^2 & xy & y^2 & x & y & 1 \\
x_1^2 & x_1y_1 & y_1^2 & x_1 & y_1 & 1 \\
x_2^2 & x_2y_2 & y_2^2 & x_2 & y_2 & 1 \\
x_3^2 & x_3y_3 & y_3^2 & x_3 & y_3 & 1 \\
x_4^2 & x_4y_4 & y_4^2 & x_4 & y_4 & 1 \\
x_5^2 & x_5y_5 & y_5^2 & x_5 & y_5 & 1
\end{vmatrix} = 0, \tag{4.27}
$$

which is obtained from (4.26) by setting

$$
\mathbf{x} = \begin{bmatrix} x \\ y \end{bmatrix} \leftarrow \begin{bmatrix} x/z \\ y/z \\ 1 \end{bmatrix}.
$$

Left of the arrow, we have affine coordinates, while there are projective ones right of it!

It should be noted that the process of finding an interpolating conic may be numerically unstable: if the five points are "close", they do not convey much information about the conic in an intuitive sense. If we use (4.26) or (4.27) to check if a point is on a conic, these expressions will in general not evaluate to zero, but to some small number. Thus we cannot check our determinants for being zero, but rather against a small tolerance ϵ. If the given points were too close to each other, many points not on the intended conic will satisfy the specified tolerance, and thus be labeled "on the conic".

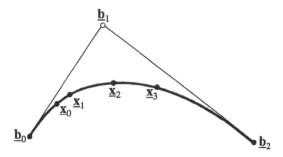

Figure 4.8. Parametric conic interpolation: the given data points together with the unknown control polygon.

4.7 Parametric Interpolating Conics

Parametric conics (4.9) may be used to solve the following interpolation problem: given are four points $\underline{\mathbf{x}}_0, \ldots, \underline{\mathbf{x}}_3$ and corresponding parameter values $t_0, \ldots, t_3$. Find a conic $\underline{\mathbf{c}}(t)$ such that $\underline{\mathbf{c}}(t_i) \hat{=} \underline{\mathbf{x}}_i$. Fig. 4.8 illustrates.

Since we will be using linear systems, we first rewrite (4.9) as

$$\underline{\mathbf{c}}(t) = [B_0^2(t), B_1^2(t), B_2^2(t)] \begin{bmatrix} \underline{\mathbf{b}}_0 \\ \underline{\mathbf{b}}_1 \\ \underline{\mathbf{b}}_2 \end{bmatrix}. \qquad (4.28)$$

Note that the column on the right has three points as its elements; thus (4.28) is really short for three dot products. We now write down the interpolation conditions for the first three data points:

$$\begin{bmatrix} z_0 \underline{\mathbf{x}}_0 \\ z_1 \underline{\mathbf{x}}_1 \\ z_2 \underline{\mathbf{x}}_2 \end{bmatrix} = \begin{bmatrix} B_0^2(t_0) & B_1^2(t_0) & B_2^2(t_0) \\ B_0^2(t_1) & B_1^2(t_1) & B_2^2(t_1) \\ B_0^2(t_2) & B_1^2(t_2) & B_2^2(t_2) \end{bmatrix} \begin{bmatrix} \underline{\mathbf{b}}_0 \\ \underline{\mathbf{b}}_1 \\ \underline{\mathbf{b}}_2 \end{bmatrix}. \qquad (4.29)$$

The factors z_i are — unknown — scaling factors. We can rewrite (4.29) by naming the involved vectors[5] and matrix:

$$\boxed{\mathbf{x}} = B \boxed{\mathbf{b}}.$$

This has the solution $\boxed{\mathbf{b}} = B^{-1} \boxed{\mathbf{x}}$. The problem is, $\boxed{\mathbf{x}}$ involves the unknown z_i, and they still have to be found. Now we make use of the

---

[5]Column vectors that have points as their elements are marked by boxes. Keep in mind that this notation is shorthand for *three* linear systems, one for each of the $x-, y-, z-$ components!

fourth point, $\underline{x}_3$. We may set $z = 1$ for one point without loss of generality, and we do this for $\underline{x}_3$. Then (4.28) becomes

$$\underline{x}_3 = \underline{c}(t_3) = [B_0^2(t_3), B_1^2(t_3), B_2^2(t_3)] \begin{bmatrix} \underline{b}_0 \\ \underline{b}_1 \\ \underline{b}_2 \end{bmatrix}. \qquad (4.30)$$

Abbreviating this as $\underline{x}_3 = B_3\boxed{\underline{b}}$, we have

$$\underline{x}_3 = B_3 B^{-1}\boxed{\underline{x}}.$$

Although it does not look like it at first sight, this actually constitutes a 3×3 linear system for the z_i, which are "hidden" in $\boxed{\underline{x}}$. There is no guarantee, of course, that the z_i will turn out being positive — thus if we project into affine space (see Chapter 5), we may encounter asymptotes.

Normally, we say that a conic is determined by five points on it. In that case, no parameter values are used. If we introduce parameters that are attached to the data points, then four points are sufficient!

4.8   Blossoms and Polars

We now introduce a generalization of the de Casteljau algorithm from Section 4.2. It will unify several of the previously developed concepts.

The de Casteljau algorithm for conics is a two-level procedure: at the first level, we interpolate (twice) with respect to t to obtain the two $\underline{b}_i^1(t)$. At the second level, we interpolate with respect to t again, obtaining $\underline{b}_0^2(t)$. Our generalization is to allow two *different* parameters at the two levels. While the "standard" de Casteljau algorithm yields a point that depends on one variable: $\underline{b}_0^2(t)$, we now obtain a point that depends on two variables. We denote it by $\underline{b}[s, t]$, the two variables being s and t. Of course, if $s = t$, we have $\underline{b}[t, t] = \underline{b}_0^2(t)$. We call $\underline{b}[s, t]$ the *blossom* of the conic; this terminology was introduced by L. Ramshaw in 1987, see [127] or [128].

The blossom $\underline{b}[s, t]$ is a *symmetric function*: this means that $\underline{b}[s, t] = \underline{b}[t, s]$. This is verified by a straightforward exercise in algebra.

Just to get used to the idea of blossoms, let us compute a few special values. We already know $\underline{b}[0, 0] = \underline{b}_0$ and $\underline{b}[1, 1] = \underline{b}_2$. Next, let us try $\underline{b}[0, 1]$. We get the following scheme:

$$\begin{array}{lll} \underline{b}_0 & & \\ \underline{b}_1 & \underline{b}_0 & \\ \underline{b}_2 & \underline{b}_1 & \underline{b}_1 = \underline{b}[0, 1], \end{array}$$

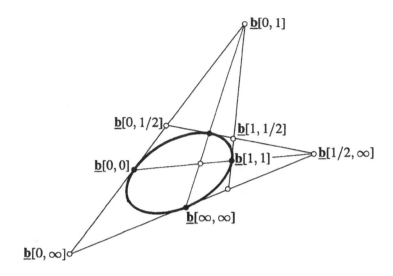

Figure 4.9. Conic blossoms: several blossom values.

where we have used $s = 0$ first and $t = 1$ second — the reader is invited to verify that the same result is obtained by using $t = 1$ first and then $s = 0$. Thus all conic control points can be written as blossom values!

Similarly, we can show that $\underline{\mathbf{b}}[0, t] = \underline{\mathbf{b}}_0^1(t)$ and $\underline{\mathbf{b}}_1^1(t) = \underline{\mathbf{b}}[1, t]$. Thus we can reformulate the de Casteljau scheme in blossom form:

$$
\begin{array}{lll}
\underline{\mathbf{b}}_0 & & \\
\underline{\mathbf{b}}_1 & \underline{\mathbf{b}}_0^1(t) & \\
\underline{\mathbf{b}}_2 & \underline{\mathbf{b}}_1^1(t) & \underline{\mathbf{b}}_0^2(t)
\end{array}
\quad = \quad
\begin{array}{lll}
\underline{\mathbf{b}}[0, 0] & & \\
\underline{\mathbf{b}}[0, 1] & \underline{\mathbf{b}}[0, t] & \\
\underline{\mathbf{b}}[1, 1] & \underline{\mathbf{b}}[1, t] & \underline{\mathbf{b}}[t, t].
\end{array}
$$

Fig. 4.9 shows a conic with several blossom values; we will now discuss those.[6] First, we observe that if r is a fixed parameter, then $\underline{\mathbf{b}}[r, t]$ traces out all points on the tangent $\underline{\mathbf{T}}(r)$ at $\underline{\mathbf{b}}[r, r]$. In particular, $\underline{\mathbf{b}}[r, t]$ is the intersection of the tangents at $\underline{\mathbf{b}}[r, r]$ and $\underline{\mathbf{b}}[t, t]$:

$$\underline{\mathbf{b}}[r, t] = \underline{\mathbf{T}}(r) \wedge \underline{\mathbf{T}}(t), \tag{4.31}$$

which gives a geometric argument for the symmetry of the blossom function.

In classical geometry, one also sees the definition that the point $\underline{\mathbf{b}}[r, t]$ is the *pole* to the *polar*

$$\underline{\mathbf{B}}[r, t] = \underline{\mathbf{b}}[r, r] \wedge \underline{\mathbf{b}}[t, t],$$

---

[6] Compare with Fig. 4.7!

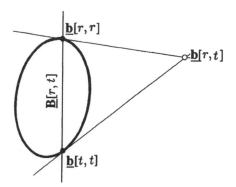

Figure 4.10. Pole and polar: a point outside the conic is related to a secant of the conic, namely its polar.

as illustrated in Fig. 4.10. Note that not every point has a corresponding polar — we call those points interior to the conic. If we wanted to be able to define polars for *every* point, we would have to work in a projective plane that is defined over complex numbers.

For the case $r = t$, this becomes

$$\underline{\mathbf{B}}[r, r] = \underline{\mathbf{T}}(r),$$

i.e., the polar of a point on a conic is its tangent.

We can also write derivatives of a conic in blossom form:

$$\dot{\underline{\mathbf{b}}}(t) \hat{=} \underline{\mathbf{b}}[\infty, t], \quad \ddot{\underline{\mathbf{b}}}(t) \hat{=} \underline{\mathbf{b}}[\infty, \infty]. \tag{4.32}$$

Thus taking derivatives amounts to evaluating at $t = \infty$, and all derivatives to a conic are on the line $\underline{\mathbf{B}}[\infty, t]$.

We note the following identity:

$$\mathrm{cr}(\underline{\mathbf{b}}[0, t], \underline{\mathbf{b}}[t, t], \underline{\mathbf{b}}[1, t], \underline{\mathbf{b}}[\infty, t]) = \frac{1}{2}, \tag{4.33}$$

i.e., the four points with parameter values $0, t, 1, \infty$ on a tangent $\underline{\mathbf{T}}(t)$ are harmonic. This was already discussed in Section 4.4.

A more general definition of poles and polars also exists (see, e.g., [144]): let $\underline{\mathbf{p}}$ be an arbitrary point. Draw any line through $\underline{\mathbf{p}}$ and record its two intersections $\underline{\mathbf{p}}_1, \underline{\mathbf{p}}_2$ with a given conic.[7] The locus of all points $\underline{\mathbf{p}}^*$ such that $\underline{\mathbf{p}}_1, \underline{\mathbf{p}}^*, \underline{\mathbf{p}}_2, \underline{\mathbf{p}}$ are harmonic is a line, namely the polar with respect to $\underline{\mathbf{p}}$.

---

[7]In order to always have two intersections, one might have to resort to working with the complex projective plane.

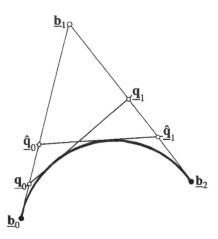

Figure 4.11. Conic reparametrization: a new tangent can be given the role of shoulder tangent.

4.9   Reparametrization

In our definition of parametric conics, we had picked a tangent through two points $\underline{\mathbf{q}}_0$ and $\underline{\mathbf{q}}_1$. We then assumed that all point coordinates were rescaled such that

$$\underline{\mathbf{q}}_0 = \frac{1}{2}\underline{\mathbf{b}}_0 + \frac{1}{2}\underline{\mathbf{b}}_1 \quad \text{and} \quad \underline{\mathbf{q}}_1 = \frac{1}{2}\underline{\mathbf{b}}_1 + \frac{1}{2}\underline{\mathbf{b}}_2 \tag{4.34}$$

Let $\hat{\underline{\mathbf{q}}}_0 \wedge \hat{\underline{\mathbf{q}}}_1$ define yet another tangent to our conic, as shown in Fig. 4.11. Suppose that, with (4.34) still holding, we have $\hat{\underline{\mathbf{q}}}_0 = \underline{\mathbf{b}}_0 + c\underline{\mathbf{b}}_1$, such that $\mathrm{cr}(\underline{\mathbf{b}}_0, \hat{\underline{\mathbf{q}}}_0, \underline{\mathbf{q}}_0, \underline{\mathbf{b}}_1) = c$. Because of the four tangent theorem, the four points $\underline{\mathbf{b}}_1, \hat{\underline{\mathbf{q}}}_1, \underline{\mathbf{q}}_1, \underline{\mathbf{b}}_2$ must also be in cross ratio c.

Following a similar development to that in Section 4.2, we arrive at

$$\underline{\mathbf{b}}(s) = (1-s)^2\underline{\mathbf{b}}_0 + 2cs(1-s)\underline{\mathbf{b}}_1 + c^2 s^2 \underline{\mathbf{b}}_2 \tag{4.35}$$

as the equation of the conic with $\hat{\underline{\mathbf{q}}}_0 \wedge \hat{\underline{\mathbf{q}}}_1$ as the shoulder tangent corresponding to $s = \frac{1}{2}$. The choice of a new shoulder tangent can be viewed as a *reparametrization* of the conic. Since

$$\hat{\underline{\mathbf{q}}}_0 = \underline{\mathbf{b}}_0 + c\underline{\mathbf{b}}_1,$$

the new parameter s is related to the old parameter t by

$$t = 0 \leftrightarrow s = 0, \quad t = \frac{1}{2} \leftrightarrow s = r, \quad t = 1 \leftrightarrow s = 1, \tag{4.36}$$

with $r = c/(1 + c)$ since then $cr(0, r, \frac{1}{2}, 1) = c$. Note that for $s = 1$, we obtain the point $c^2 \mathbf{b}_2$ on the reparametrized curve — equivalent of course to $\underline{\mathbf{b}}_2$.

Both t and s represent parametrizations of the real line; they are related by a Moebius transformation as described in Section 2.4. It is characterized by

$$cr\left(0, t, \frac{1}{2}, 1\right) = cr(0, s, r, 1)$$

and thus

$$s = \frac{ct}{1 + (c - 1)t}. \tag{4.37}$$

It is important to note that, although the shape of the conic is unchanged by such a reparametrization, its derivatives change: we now have

$$\underline{\dot{\mathbf{b}}}(0) = 2c\underline{\mathbf{b}}_1 - 2\underline{\mathbf{b}}_0 \quad \text{and} \quad \underline{\dot{\mathbf{b}}}(1) = 2c^2\underline{\mathbf{b}}_2 - 2c\underline{\mathbf{b}}_1. \tag{4.38}$$

Using the concept of reparametrization, we may bring conics into *standard form*. Let some conic have weights z_0, z_1, z_2; assume that $z_0 = 1$ (thus excluding the case $z_0 = 0$!). We can now find a new set of weights $\hat{z}_i$, namely $1, \hat{z}_1, 1$ by setting

$$c = \sqrt{\frac{1}{z_2}}. \tag{4.39}$$

This does indeed achieve $\hat{z}_2 = 1$, as you might want to check! Of course, this only works if $z_2 > 0$ — thus not all quadratics can be brought into standard form. In Chapter 5, we will mostly be concerned with conics for which all three z_i are positive; then the standard form is always possible.

4.10   The Complementary Segment

Let us consider the particular reparametrization arising from selecting $c = -1$ in (4.35). Then the three points $\underline{\mathbf{b}}(t), \underline{\mathbf{b}}(s), \mathbf{b}_1$ are collinear, as follows from

$$\begin{aligned}
\det[\underline{\mathbf{b}}(t),\ \underline{\mathbf{b}}(s),\ \mathbf{b}_1] &= \det\left[\mathbf{b}_0 B_0^2(t) + \mathbf{b}_1 B_1^2(t) + \mathbf{b}_2 B_2^2(t),\ \mathbf{b}_0 B_0^2(t)\right. \\
&\qquad \left. -\mathbf{b}_1 B_1^2(t) + \mathbf{b}_2 B_2^2(t),\ \mathbf{b}_1\right] \\
&= \det[\underline{\mathbf{b}}_1(B_1^2(t) - B_1^2(s)),\ \underline{\mathbf{b}}(s),\ \mathbf{b}_1] \\
&= 0,
\end{aligned}$$

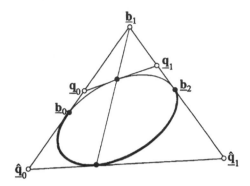

Figure 4.12. The complementary segment: reversing the weight of the middle control point generates the "other" part of the conic.

assuming standard form for both conics. Referring to Fig. 4.12, we see that $\underline{\mathbf{b}}(s), 0 \le s \le 1$, traces out the part of the conic that is "missed" by $\underline{\mathbf{b}}(t), 0 \le t \le 1$.

A different way of seeing this is provided by the following argument: instead of the tangent $\underline{\mathbf{T}}(\frac{1}{2}) = \underline{\mathbf{q}}_0 \wedge \underline{\mathbf{q}}_1$, we might have prescribed $\underline{\mathbf{T}}(\infty)$ as the shoulder tangent. The roles of $\underline{\mathbf{T}}(\frac{1}{2})$ and $\underline{\mathbf{T}}(\infty)$ are then reversed.

A cautionary note: the correct depiction of the geometry from Fig. 4.12 looks like Fig. 4.13: since $\hat{z}_1 < 0$, the control polygon crosses infinity!

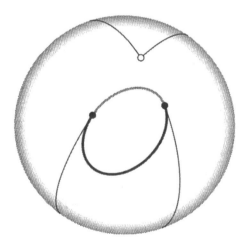

Figure 4.13. The complementary segment: the actual geometry.

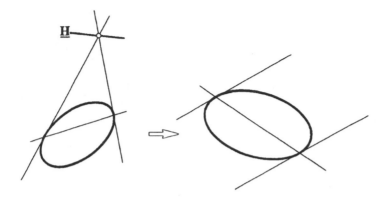

Figure 4.14. Conjugate tangents and secants: if a projective pole is mapped to $\mathbf{L}_\infty$, its corresponding tangents become parallel.

4.11 Parametric Affine Conics

If we project from $I\!\!P^2$ into the extended affine plane, conics take on the familiar shapes of ellipses, hyperbolas, and parabolas, as discussed in Section 3.5. Let $\underline{\mathbf{H}}$ (for *horizon*) be the pre-image of $\mathbf{L}_\infty$. Let us assume for a moment that $\underline{\mathbf{H}}$ does not intersect the conic. Let $\underline{\mathbf{b}}[r,s]$ on $\underline{\mathbf{H}}$ be the pole to the polar $\underline{\mathbf{B}}[r,s]$. The affine images of $\underline{\mathbf{B}}[r,r]$ and $\underline{\mathbf{B}}[s,s]$ will be two *parallel tangents*, while $\underline{\mathbf{B}}[r,s]$ will be mapped to a secant of the affine conic. We call the tangents $\underline{\mathbf{B}}[r,r]$ and $\underline{\mathbf{B}}[s,s]$ *conjugate* to the secant $\underline{\mathbf{B}}[r,s]$. See Fig. 4.14 for an illustration.

4.12 Problems

1. Show that the point $\underline{\mathbf{b}}_0^2$ defined by (4.9) does in effect lie on the tangent $\underline{\mathbf{T}}$ defined by (4.10), i.e., show that

$$\underline{\mathbf{T}}\underline{\mathbf{b}}_0^2 = 0.$$

2. Show that the implicit form of an affine conic is given by

$$\mathbf{x}^\mathrm{T} A\mathbf{x} + 2\mathbf{b}^\mathrm{T}\mathbf{x} + d = 0, \tag{4.40}$$

with $\mathbf{x} = [x,y]^\mathrm{T}$, A a 2×2 symmetric matrix, and $\mathbf{b} = [b,c]^\mathrm{T}$. Hint: use (4.6).

3. In Section 4.7, there was no requirement that the parameter values t_i be increasing. Investigate what will happen if, for the configuration of Fig. 4.8, the t_i are given in nondecreasing order. Example: 0.2, 0.6, 0.4, 0.8.

4. Use the concept of the complementary segment to prove the special case of Brianchon's theorem as shown in Fig. 3.14.

5. A point conic is the set of all points which satisfy $\mathbf{x}^T A \mathbf{x} = 0$. If we write this conic as a line conic, we view it as a set of lines $\underline{\mathbf{T}}$ which are tangent to the conic. Show that the implicit equation in line form is given by

$$\underline{\mathbf{T}} A^{-1} \underline{\mathbf{T}}^T = 0. \tag{4.41}$$

5

Rational Quadratic Conics

While the projective treatment of conics reveals most of their fundamental properties in an elegant way, "real world" applications of conics happen in an affine environment. In this chapter, we will thus shift our emphasis from the projective treatment of conics to an affine viewpoint. [1]

5.1 The Rational Quadratic Form

In the projective plane $I\!P^2$, a conic may be described by a quadratic Bézier curve:

$$
\begin{aligned}
\underline{\mathbf{b}}^2(t) &= (1-t)^2\underline{\mathbf{b}}_0 + 2t(1-t)\underline{\mathbf{b}}_1 + t^2\underline{\mathbf{b}}_2 \\[2mm]
&= (1-t)^2 \begin{bmatrix} x_0 \\ y_0 \\ z_0 \end{bmatrix} + 2t(1-t) \begin{bmatrix} x_1 \\ y_1 \\ z_1 \end{bmatrix} + t^2 \begin{bmatrix} x_2 \\ y_2 \\ z_2 \end{bmatrix} \\[2mm]
&= (1-t)^2 z_0 \begin{bmatrix} \mathbf{b}_0 \\ 1 \end{bmatrix} + 2t(1-t) z_1 \begin{bmatrix} \mathbf{b}_1 \\ 1 \end{bmatrix} + t^2 z_2 \begin{bmatrix} \mathbf{b}_2 \\ 1 \end{bmatrix},
\end{aligned}
$$

---

[1] Here, we are dealing exclusively with planar curves. Hence we denote the homogeneous coordinate by z, as opposed to the homogeneous w in subsequent chapters.

81

Figure 5.1. Weight signs: for $z_0 > 0$ and $z_1, z_2 < 0$, the weight function has a zero in $[0, 1]$.

where the $\mathbf{b}_i$ are already written as affine points:

$$\mathbf{b}_i = \frac{1}{z_i} \begin{bmatrix} x_i \\ y_i \end{bmatrix}. \tag{5.1}$$

We can now project our projective curve $\underline{\mathbf{b}}^2(t)$ into the (extended) affine plane, yielding the *rational quadratic form of a conic:*

$$\mathbf{b}^2(t) = \frac{(1-t)^2 z_0 \mathbf{b}_0 + 2t(1-t)z_1 \mathbf{b}_1 + t^2 z_2 \mathbf{b}_2}{(1-t)^2 z_0 + 2t(1-t)z_1 + t^2 z_2}. \tag{5.2}$$

The (affine) points $\mathbf{b}_i$ are called *control points*, and the z_i are called *weights*.

The projective shoulder tangent $\underline{\mathbf{T}}(\frac{1}{2})$ is mapped to the tangent $\mathbf{T}(\frac{1}{2})$ of the affine conic; the points $\underline{\mathbf{q}}_i$ are mapped to affine points $\mathbf{q}_i$:

$$\mathbf{q}_0 = \frac{z_0 \mathbf{b}_0 + z_1 \mathbf{b}_1}{z_0 + z_1}, \quad \mathbf{q}_1 = \frac{z_1 \mathbf{b}_1 + z_2 \mathbf{b}_2}{z_1 + z_2}. \tag{5.3}$$

Because of their dependence on the weights, we call the $\mathbf{q}_i$ *weight points*. The relation between weights and weight points is given by

$$\text{ratio}(\mathbf{b}_0, \mathbf{q}_0, \mathbf{b}_1) = \frac{z_1}{z_0} \quad \text{and} \quad \text{ratio}(\mathbf{b}_1, \mathbf{q}_1, \mathbf{b}_2) = \frac{z_2}{z_1}.$$

For the definition of ratios, see Section 2.7.

In the affine case, the *signs* of the weights are of considerable importance; we will require that they are all of the same sign. To see why this is desirable, let us investigate the case of z_0 being negative, while z_1 and z_2 are positive. Referring to Fig. 5.1, we see that the denominator of (5.2), $z(t) = (1-t)^2 z_0 + 2t(1-t)z_1 + t^2 z_2$, the *weight function*, has a zero t^* in the interval $[0, 1]$. Assuming finite $\mathbf{b}_i$, this will cause the conic to have a singularity, i.e., it will cross the line at infinity. The tangent at t^* is an

Figure 5.2. Weight signs: for $z_0 > 0$ and $z_1, z_2 < 0$, the conic has a singularity for $t^* \in [0, 1]$.

asymptote of the conic, which consequently is a hyperbola, as discussed in Section 3.5. Fig. 5.2 shows the behavior of both curve and control polygon — not very desirable for "real life" situations! Note that the control polygon itself crosses the horizon because of the weight signs.

Unless otherwise stated, we will thus assume that all weights are positive,[2] even at the expense of leaving out hyperbolas with an asymptote in the interval $[0, 1]$. From Section 4.9, we know that it is then no restriction to assume that $z_0 = z_2 = 1$, and we will also assume that conics are given in this *standard form* from now on.

5.2 Conic Classification

An affine conic in standard form has a remarkable geometric property: its shoulder tangent is parallel to the chord $\overline{\mathbf{b}_0 \mathbf{b}_2}$. This follows from the observation that

$$\text{ratio}(\mathbf{b}_0, \mathbf{q}_0, \mathbf{b}_1) = \text{ratio}(\mathbf{b}_2, \mathbf{q}_1, \mathbf{b}_1) = z_1.$$

Let us again denote by $\underline{\mathbf{H}}$ the projective line that corresponds to the affine line at infinity. Then we know that the projective lines $\underline{\mathbf{q}}_0 \wedge \underline{\mathbf{q}}_1$ and $\underline{\mathbf{b}}_0 \wedge \underline{\mathbf{b}}_2$ intersect on $\underline{\mathbf{H}}$, since that is the condition for them to be mapped to parallel lines in affine space. So if we standardize an affine conic by selecting a

---

[2]They may also be all negative — a common factor of -1 does not change anything!

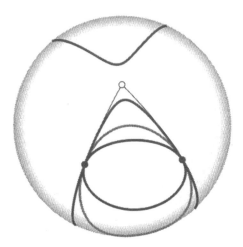

Figure 5.3. Conic classification: in the extended affine plane, three types of conics exist. The parabola is shown in gray.

shoulder tangent parallel to $\overline{\mathbf{b}_0\mathbf{b}_2}$, this corresponds to choosing a projective shoulder tangent that passes through $[\underline{\mathbf{b}}_0 \wedge \underline{\mathbf{b}}_2] \wedge \underline{\mathbf{H}}$. In Fig. 5.3, we utilize the extended affine plane: the horizon corresponds to $\underline{\mathbf{H}}$. The three affine conic types are shown: the ellipse is a finite curve, the parabola (gray) touches the horizon, and the hyperbola intersects it twice.

Referring to Fig. 4.7, we can also say that a *parabola* in standard form goes to infinity for the value $t = \pm\infty$. The situation for a *hyperbola* in standard form is more involved; see Fig. 5.4.

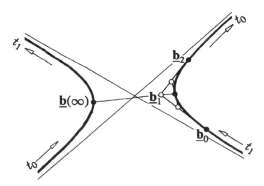

Figure 5.4. Hyperbolas in standard form: the figure shows how the parameter t traverses the curve.

Figure 5.5. Conic classification: the zeroes of the weight function determine the type of conic. From left to right: weight functions for a hyperbola, a parabola, and an ellipse.

The weight function $z(t)$ provides us with a more algebraic means to classify rational quadratic conics into the three affine types: ellipses, hyperbolas, and parabolas. The singularities of the conic, i.e., the points where it meets the line at infinity, correspond to the zeroes of $z(t)$. These zeroes, if they exist, are outside the interval $[0, 1]$, and Fig. 5.5 shows the three possibilities:

- either $z_1 < 1$, then there are no zeroes, and the conic is an ellipse,

- or $z_1 > 1$, then there are two finite zeroes, symmetric with respect to $t = 1/2$, and the conic is a hyperbola,

- or $z_1 = 1$, then there are two infinite zeroes, and the conic is a parabola.

The counting of the zeroes for the parabola weight function should be interpreted as the transition going from the hyperbolic to the elliptic case: as z_1 approaches 1 from above or from below, the two zeroes tend to infinity.

5.3 Properties of Rational Quadratics

In this section, we list some properties exhibited by all conics in rational quadratic form; properties of the different types are discussed at the end of this chapter. We assume that the conics are given in standard form.

Convex hull property: The conic segment corresponding to $t \in [0, 1]$ lies in the *convex hull* of the control polygon, i.e., it is contained inside the triangle formed by the control polygon. A stronger statement can be made also: the curve segment is contained inside the quadrilateral $\mathbf{b}_0, \mathbf{q}_0, \mathbf{q}_1, \mathbf{b}_2$, where the $\mathbf{q}_i$ are the weight points from (5.3).

Affine invariance: The following two procedures yield the same result. First, evaluate $\mathbf{b}(t)$ and map it to $\Phi\mathbf{b}(t)$ by an affine map (see Section 2.7). Second, map all $\mathbf{b}_i$ to $\Phi\mathbf{b}_i$ (leaving the weight z_1 unchanged) and evaluate the new rational quadratic at t. Note that the type of conic is not changed by an affine map.

Projective invariance: In the same sense, rational quadratics are projectively invariant. Note, however, that the ratios ratio$(\mathbf{b}_i, \mathbf{q}_i, \mathbf{b}_{i+1})$ and thus the weights will be changed after application of a projective map. Hence also the type (ellipse etc.) of the conic may be changed.

Subdivision: The shoulder point $\mathbf{b}(\frac{1}{2})$ subdivides the curve into two pieces, a "left" and a "right" one. The control polygon for the left segment is given by $\mathbf{b}_0, \mathbf{q}_0 = \mathbf{b}_0^1(\frac{1}{2}), \mathbf{b}(\frac{1}{2})$, with weights $1, (1+z_1)/2, (1+z_1)/2$. The right segment is treated analogously. Note that neither segment's weights are now in standard form.

5.4   Derivatives

Differentiation of conics is easy in projective space: there, they are quadratic polynomials, and their derivatives are linear, see Section 4.4. In affine space, however, we are dealing with rational quadratic polynomials, which would need the quotient rule for taking derivatives. A simple trick avoids this unpleasant task: our conic $\mathbf{c}(t)$ is of the form $\mathbf{c}(t) = \mathbf{p}(t)/z(t)$, with a polynomial numerator $\mathbf{p}(t)$. Thus

$$\mathbf{p}(t) = z(t)\mathbf{c}(t).$$

This polynomial curve is differentiated using the product rule:

$$\dot{\mathbf{p}}(t) = \dot{z}(t)\mathbf{c}(t) + z(t)\dot{\mathbf{c}}(t),$$

the dot denoting differentiation with respect to t. The expression $\dot{\mathbf{c}}(t)$ on the right hand side is our desired conic derivative! We can solve for it and obtain

$$\dot{\mathbf{c}}(t) = \frac{1}{z(t)}[\dot{\mathbf{p}}(t) - \dot{z}(t)\mathbf{c}(t)]. \tag{5.4}$$

This procedure may be repeated to yield higher order derivatives; see Section 7.8 for details.

We may evaluate (5.4) at the endpoints $t = 0$ and $t = 1$:

$$\dot{\mathbf{c}}(0) = \frac{2z_1}{z_0}\Delta\mathbf{b}_0, \qquad \dot{\mathbf{c}}(1) = \frac{2z_1}{z_2}\Delta\mathbf{b}_1. \tag{5.5}$$

Some applications, such as curvature computations, also require second derivatives. We obtain

$$\ddot{\mathbf{c}}(0) = \frac{1}{z_0^4}[\Delta^2(z_0\mathbf{b}_0) - z_0\mathbf{b}_0\Delta^2 z_0]\ddot{\mathbf{c}}(1) = \frac{1}{z_2^4}[\Delta^2(z_0\mathbf{b}_0) - z_2\mathbf{b}_2\Delta^2 z_0]. \tag{5.6}$$

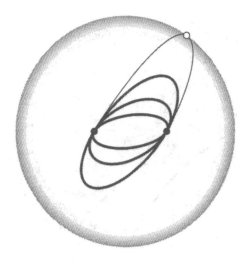

Figure 5.6. Control vectors: a conic with a control vector b_1. The effect of scaling b_1's coordinates is also shown.

5.5   Control Vectors

It is possible that one or two of the control points $\underline{b}_i$ of the projective conic are on $\underline{H}$, the pre-image of the affine line at infinity. They are then mapped to *control vectors*. Fig. 5.7 gives an example. Varying the length of a control vector has a predictable effect on the conic, as illustrated in Fig. 5.6.

Suppose that b_1 is a control vector while b_0 and b_2 are control points; thus $\underline{b}_1$'s third component z_1 is zero. We can then rewrite the rational form of this conic:

$$b(t) = \frac{z_0 b_0 B_0^2(t) + z_2 b_2 B_2^2(t)}{z_0 B_0^2(t) + z_2 B_2^2(t)} + \frac{b_1 B_1^2(t)}{z_0 B_0^2(t) + z_2 B_2^2(t)}.$$

For a more general treatment, see Section 7.11

Control vectors were first considered by K. Vesprille [151], and later by L. Piegl [118]. Both authors use the term "infinite control points". As we have seen, an infinite point (short for a point at infinity) corresponds to a whole family of parallel vectors. Each vector in this family gives rise to a different curve, but corresponds to the same point at infinity. The "infinite control point" terminology is thus oversimplifying.[3] Fiorot and Jeannin [68] share this criticism.

---

[3]T. DeRose brought this point to my attention.

Figure 5.7. Control vectors: a hyperbola with two control vectors $\mathbf{b}_0, \mathbf{b}_2$ and a control point $\mathbf{b}_1$.

If the representation of a conic involves control vectors, the convex hull property is lost — but the use of the weight points $\mathbf{q}_i$ recovers it: the curve segment is still inside the quadrilateral $\mathbf{b}_0, \mathbf{q}_0, \mathbf{q}_1, \mathbf{b}_2$.

It is also possible that $\underline{\mathbf{b}}_0$ and $\underline{\mathbf{b}}_2$ are on $\underline{\mathbf{H}}$. Then, the resulting affine conic crosses the line at infinity twice, and thus is a hyperbola. Fig. 5.7 illustrates.

Finally, a practical note on control vectors: the current standard format for geometry definition in CAD/CAM systems, IGES[4] does not allow zero weights, see also Section 15.1. This means that data exchange will be impossible if IGES is chosen as the neutral data format and control vectors are involved. Some systems try to circumvent this by replacing the zero weights by very small constants, like 10^{-10}.

5.6 The Parabola

The parabola enjoys a special role among conics in rational quadratic form: it allows a polynomial, or integral form. In that form, its weight points are simply the midpoints of the control polygon legs. The de Casteljau algorithm simplifies since we can use ratios instead of cross ratios:

$$\mathrm{cr}(\mathbf{b}_i^r, \mathbf{b}_i^{r+1}, \mathbf{q}_i^r, \mathbf{b}_{i+1}^r) = \mathrm{ratio}(\mathbf{b}_i^r, \mathbf{b}_i^{r+1}, \mathbf{b}_{i+1}^r) = \frac{t}{1-t}, \quad r = 1, 2; \ i = 0, 2-r$$

---

[4]IGES is short for Initial Graphics Exchange Standard.

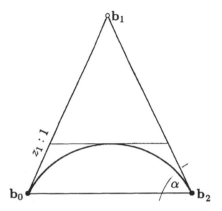

Figure 5.8. The circle: the geometry of the control polygon.

and the de Casteljau algorithm becomes

$$\mathbf{b}_i^{r+1}(t) = (1 - t)\mathbf{b}_i^r(t) + t\mathbf{b}_{i+1}^r(t). \tag{5.7}$$

But we can also write a parabola as a true rational quadratic: employing the technique of *reparametrization* from Section 4.9, we can write it as

$$\mathbf{b}(t) = \frac{\mathbf{b}_0 B_0^2(t) + c\mathbf{b}_1 B_1^2(t) + c^2\mathbf{b}_2 B_2^2(t)}{B_0^2(t) + cB_1^2(t) + c^2 B_2^2(t)} \tag{5.8}$$

with an arbitrary nonzero constant c. Only for $c = 1$, we obtain the polynomial form. For $c \neq 1$, the point $\mathbf{b}(\infty)$ will be a finite point!

5.7   The Circle

Of all conics, the circular arc is the most widely used one. Its control polygon must satisfy a special condition: it has to form an isosceles triangle, due to the circle's symmetry properties. Referring to Fig. 5.8, and assuming standard form, recall that $z_1 = \mathrm{ratio}(\mathbf{b}_0, \mathbf{q}_0, \mathbf{b}_1)$. Thus

$$z_1 = \cos \alpha,$$

with $\alpha = \angle(\mathbf{b}_2, \mathbf{b}_0, \mathbf{b}_1)$.

In the case $\alpha = 90°$, the weight vanishes and we have the control vector form of a conic, as outlined in Section 5.5.

A whole circle may be represented in many ways by piecewise rational quadratics. One example is to use the control vector form to write the "top" part, and then use the complementary segment from Section 4.10 to

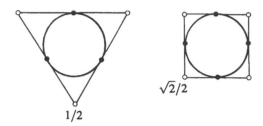

Figure 5.9. The full circle: it may be represented by three or four rational quadratics, with the weights of the control points indicated.

write the "bottom" part. It is probably more convenient — retaining the convex hull property for positive weights — to dissect the full circle into three or four parts, as shown in Fig. 5.9.

Although we can write an arc of a circle in rational quadratic form, one should not overlook that we then sacrifice one nice property of the familiar sin/cos parametrization: in the rational quadratic form, the parameter t does not traverse the circle with *unit speed*. If a curve is traversed with unit speed, its parameter is the *arc length* parameter. Unit speed traversal of a curve means that the tangent vector $\dot{\mathbf{x}}(t)$ is of unit length.[5] Clearly, the derivative of a rational quadratic does not have this property. Thus, if an arc of a rational quadratic is to be split into a certain number of segments, each subtending the same angle, numerical techniques must be invoked. In the sin/cos parametrization, by contrast, equal parameter increments ensure equal subtended angles.

It can be shown that higher degree rational curves also do not have a constant length first derivative vector, see [64].

Circular arcs may be pieced together to form *circle splines*, with easy control of tangent continuity. These splines play a role in numerically controlled (NC) manufacturing, as some NC machines can handle circles in hardware. The literature on circle splines includes: [28], [113], [132], [141].

5.8   Problems

1. In Fig. 5.7, define a coordinate system with origin $\mathbf{b}_1$ and axis directions $\mathbf{b}_0$ and $\mathbf{b}_2$. Show that in this coordinate system, the hyperbola takes on the familiar form $y = 1/x$.

---

[5]It will be of constant length if we are dealing with circles other than the unit circle.

2. A conic may have both $\underline{q}_0$ and $\underline{q}_1$ on the horizon $\underline{\mathbf{H}}$. Illustrate the resulting affine configuration.

3. Find the reparametrization $\phi = \phi(t)$ of (10.2) such that we obtain the circle equation $[x(\phi), y(\phi)]^T = [\cos(\phi), \sin(\phi)]^T$.

6

Conic Splines

Rational quadratics may be "pieced together", so as to form composite curves that can be much more complex than single conics can be. Such curves are called *conic splines*. Where two pieces, or *segments*, of a conic spline meet, the curve will be of a certain smoothness: it could be analytically differentiable: C^1, C^2, or it could be geometrically smooth: G^1, G^2. The different cases are discussed in this chapter.

6.1 C^1 Conditions

Conics may be pieced together to form a more complicated curve, whose shape is too complex to be captured by a single conic. Such composite curves are called *conic splines*. When projected into affine space, it will be an instance of NURB curves.

Fig. 6.1 shows two conics: one with control polygon $\underline{\mathbf{b}}_0, \underline{\mathbf{b}}_1, \underline{\mathbf{b}}_2$ and one with control polygon $\underline{\mathbf{b}}_2, \underline{\mathbf{b}}_3, \underline{\mathbf{b}}_4$. At $\underline{\mathbf{b}}_2$, they share a common tangent, i.e. $\det[\underline{\mathbf{b}}_1, \underline{\mathbf{b}}_2, \underline{\mathbf{b}}_3] = 0$.

Does this mean that they are C^1? In other words, do both conics share the same first derivative at $\underline{\mathbf{b}}_2$? Recall that the tangent to a curve is a line, whereas the derivative is a point, so it is not obvious that tangent continuity implies derivative continuity. Derivative continuity is a property of the

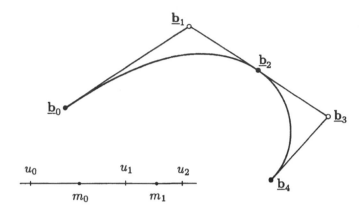

Figure 6.1. Piecewise conics: two conics form a smooth composite curve.

parametrization of a curve, i.e., more than just a geometrical property. In order to formalize this, let each segment of our composite curve be defined over an interval[1] of $\mathbb{P}^1$: let the "left" conic be defined over $[u_0, u_1]$ and the "right" conic over $[u_1, u_2]$. Let us assume a point m_i in each of the two intervals such that they are equivalent reference frames, see Section 1.4. This means that the m_i have to satisfy

$$\mathrm{cr}(u_0, m_0, u_1, \infty) = \mathrm{cr}(u_1, m_1, u_2, \infty).$$

See (2.8) for a definition of this formulation of cross products. We may thus always assume

$$m_0 = \frac{1}{2}u_0 + \frac{1}{2}u_1 \ \ \text{and} m_1 = \frac{1}{2}u_1 + \frac{1}{2}u_2.$$

If we are dealing with the real line as our domain instead of the projective one, then this simply means that the m_i are the midpoints of their respective intervals.

We now say the two segments are C^1 if the two triples $\underline{\mathbf{b}}_1, \underline{\mathbf{q}}_1, \underline{\mathbf{b}}_2$ and $\underline{\mathbf{b}}_2, \underline{\mathbf{q}}_2, \underline{\mathbf{b}}_3$ form two equivalent reference frames for the curve's tangent at $\underline{\mathbf{b}}_2$ and if these frames are a projective map of the two equivalent domain

---

[1]We assume that $\mathbb{P}^1$ is endowed with a projective reference frame, such that each point on it corresponds to a real number, or parameter. An interval $[a, b]$ of $\mathbb{P}^1$ is then the set of all points whose parameter u satisfies $a \le u \le b$. We will also refer to u as a point in $\mathbb{P}^1$. Note that this interval, when projected into affine space, may contain the point at infinity.

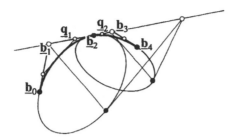

Figure 6.2. C^1 continuity: two conics form a differentiable piecewise curve at $\underline{\mathbf{b}}_2$ if the corresponding tangent has the same intersection with both conics' tangents at infinity.

frames u_0, m_0, u_1 and u_1, m_1, u_2. Since in the domain we have, with $\Delta_i = u_{i+1} - u_i$,

$$u_1 = \frac{\Delta_1}{\Delta_0 + \Delta_1} u_0 + \frac{\Delta_0}{\Delta_0 + \Delta_1} u_2,$$

the same linear relationship must hold for the points on the tangent:

$$\underline{\mathbf{b}}_2 = \frac{\Delta_1}{\Delta_0 + \Delta_1} \underline{\mathbf{b}}_1 + \frac{\Delta_1}{\Delta_0 + \Delta_1} \underline{\mathbf{b}}_3. \tag{6.1}$$

This is the condition for the left and right segment to be C^1 with respect to the *knot sequence* u_0, u_1, u_2.

Since the first derivative of a conic may be defined in terms of a blossom value at infinity (see Section 4.8), these conditions are equivalent to

$$\underline{\mathbf{b}}_{\text{left}}[u_1, \infty] \hat{=} \underline{\mathbf{b}}_{\text{right}}[u_1, \infty]. \tag{6.2}$$

This interpretation of C^1 continuity is illustrated in Fig. 6.2.

Next, we might want to ask about C^2 continuity of our composite curve.[2] The algebraic condition is that the second derivatives of both curves match:

$$\underline{\mathbf{b}}_{\text{left}}[\infty, \infty] \hat{=} \underline{\mathbf{b}}_{\text{right}}[\infty, \infty]. \tag{6.3}$$

Thus our conics, or quadratic curves, agree in position, first and second derivative at u_1. It follows that both conics have to be part of *one* conic.

An interesting consequence is that C^1 continuity between two conics implies C^2 continuity provided that they are both part of the same conic.

---

[2] When we talk about C^2 continuity, we always assume that the curve is already C^1 and C^0.

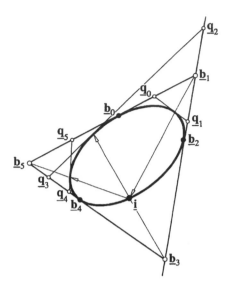

Figure 6.3. C^2 conics: a conic is covered by three quadratic segments.

Fig. 6.3 shows a conic that is covered by three C^2 quadratics. Note how each conic has the same point $\underline{\mathbf{i}}$ corresponding to $u = \infty$. In that figure, the points $\underline{\mathbf{q}}_2$ and $\underline{\mathbf{q}}_3$ are "outside" the polygon legs $\underline{\mathbf{b}}_2, \underline{\mathbf{b}}_3$ and $\underline{\mathbf{b}}_3, \underline{\mathbf{b}}_4$, respectively. When projecting into affine space, this will result in negative weights.

6.2 Parametric Continuity for Rational Conic Splines

In affine space, conics are represented by rational quadratics, and conic splines are piecewise rational quadratics. The above C^1 and C^2 conditions still hold, as derivative continuity is preserved by projecting from projective into affine space. But there are more general continuity conditions for rational splines: we may use the *quotient rule* or (5.4) for differentiating each segment! Assuming the "left" conic segment, defined over the interval $[u_0, u_1]$ has the control polygon $\mathbf{b}_0, \mathbf{b}_1, \mathbf{b}_2$, and the "right" segment, defined over $[u_1, u_2]$ is given by $\mathbf{b}_1, \mathbf{b}_2, \mathbf{b}_3$. Also assume that each control point $\mathbf{b}_i$ has a weight z_i attached to it.

The two rational quadratic conics form a C^1 curve over the domain $[u_0, u_2]$ if

$$\frac{z_1 \Delta \mathbf{b}_1}{\Delta_0} = \frac{z_3 \Delta \mathbf{b}_2}{\Delta_1}, \tag{6.4}$$

a straightforward consequence of (6.1). This condition is more general than the one obtained by demanding that both segment form a C^1 curve in projective space: that would result in the stricter requirement of C^1 continuity of each component function $x(u), y(u), z(u)$.

6.3   Conic B-Splines

When two conics meet such that they satisfy the strict C^1 condition (6.2) we know that each control point $\underline{\mathbf{b}}_{2i}$ is determined from (6.2), as is each weight z_{2i}:

$$\underline{\mathbf{b}}_{2i} = \frac{\Delta_i \underline{\mathbf{b}}_{2i-1} + \Delta_{i-1} \underline{\mathbf{b}}_{2i+1}}{\Delta_{i-1} + \Delta_i}. \tag{6.5}$$

In order to *store* this spline, its thus seems sufficient to store only the $\underline{\mathbf{b}}_{2i\pm1}$ instead of all control points. In doing so, we will relabel then as $\underline{\mathbf{d}}_i$, see Fig. 6.4 We now also make the transition to *rational splines*, as this is closer to the widely accepted IGES format. Thus instead of dealing with projective points $\underline{\mathbf{d}}_i$, we are dealing with affine points $\mathbf{d}_i$. These are obtained by dividing the first two coordinates of each $\underline{\mathbf{d}}_i$ by its third coordinate v_i. The v_i are then stored separately and are called *weights*.

A rational conic spline could be defined as follows:

1. the number N of conic segments,

2. by a sequence of $N + 2$ control points $\mathbf{d}_0, \dots, \mathbf{d}_{N+1}$,

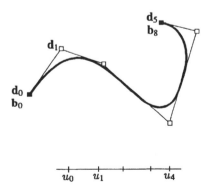

Figure 6.4. Conic B-splines: control polygon, curve, and knot sequence.

Figure 6.5. Conic splines: a curve with three conic segments (dark), together with the completed conics (gray).

3. by a sequence of weights $v_0, \ldots, v_{N+1}$,

4. by a sequence of knots $u_0, \ldots, u_N$.

The last item, the *knots*, needs some explanation. We assume that the ith conic segment is defined over the interval $[u_i, u_{i+1}]$. The increasing sequence $\{u_i\}$ of reals is called the *knot sequence*. It forms a partition of the domain $[u_0, u_N]$ of the conic spline $\mathbf{s}(u)$.

Each segment of the spline is a conic; they are pieced together so that a smooth curve results. Fig. 6.5 shows these complete conics.

The term B-spline will be explained shortly; for now, note that we have reduced the amount of storage compared to the previous data specification. Also, we are guaranteed to represent curves that adhere to the strict C^1 condition (6.2), as shown in Fig. 6.4. This data specification is essentially what we will encounter in Section 15.1 as the IGES format.

Note that conic splines that merely satisfy the weaker C^1 condition (6.4) do not allow such a compact representation: the weights z_{2i} are not tied to their neighboring weights by any condition — only the control points $\mathbf{b}_{2i}$ are![3]

Now to the term "B-spline", which is short for "basis spline". It is justified by the fact that we may write a strictly C^1 conic spline curve in

---

[3]In terms of the IGES format from Section 15.1, this means that all u_i are treated as *double knots*.

this form:

$$s(u) = \frac{\sum_{i=0}^{N+2} v_i d_i N_i^2(u)}{\sum_{i=0}^{N+2} v_i N_i^2(u)}. \tag{6.6}$$

The functions N_i^2 are the quadratic B-splines that will be introduced with in Chapter 9.

6.4   Curvature Continuity

The most significant geometric feature of any planar curve in Euclidean space is its *curvature*, or the rate of change of its tangent direction. We refer the reader to [26], [46], or [57] for the necessary differential geometry. The curvature $\kappa(t)$ of a parametric curve $\mathbf{x}(t)$ in $I\!\!E^2$ is given by

$$\kappa(t) = \frac{||\dot{\mathbf{x}} \wedge \ddot{\mathbf{x}}||}{||\dot{\mathbf{x}}||^3}. \tag{6.7}$$

The curvature is the inverse of the radius of the *osculating circle*, which is the circle that best approximates a curve at a given point. Where the curve is bent sharply, its curvature is high; where the curve is flat, its curvature is low. Note that curvature is a Euclidean concept; it cannot be defined in a straightforward manner in projective space. The curvature of a curve changes under affine and projective transformations; only (Euclidean) rotations and translations — so-called rigid body motions — leave it unchanged.[4]

Let us now try to evaluate a conic's curvature at the endpoints of its control polygon.[5] We assume the conic is in standard form; inserting the appropriate quantities for the derivatives of a rational quadratic into (6.7), and evaluating at $t = 0$ and $t = 1$, we obtain

$$\kappa(0) = \frac{\mathcal{A}}{z_1^2 \ell_0^3}, \quad \kappa(1) = \frac{\mathcal{A}}{z_1^2 \ell_1^3}, \tag{6.8}$$

where $\mathcal{A}$ denotes the area of the triangle formed by the conic control polygon, and ℓ_0, ℓ_1 are the polygon leg lengths, as shown in Fig. 6.6. Note that we can assign a *sign* to the curvature, depending on the sign of $\mathcal{A}$. See Fig. 6.6 for an illustration.

---

[4]But curvature *continuity* is invariant under all these transformations!

[5]When talking about curvature, we always refer to conics in Euclidean space — projective concepts are of little use here!

Figure 6.6. Conic curvatures: the curvature of a conic at its endpoints is determined by the shown geometric elements.

In addition to the endpoints, it is also easy to compute the curvature $\kappa(\frac{1}{2})$ of the shoulder point $\mathbf{b}(\frac{1}{2})$. It is given by

$$\kappa(\frac{1}{2}) = 8z_1 \frac{A}{||\mathbf{b}_2 - \mathbf{b}_0||^3}.$$ (6.9)

6.5 G^2 Conic Splines

Two coplanar conics in standard form, with control polygons $\mathbf{b}_0, \mathbf{b}_1, \mathbf{b}_2$ and $\mathbf{b}_2, \mathbf{b}_3, \mathbf{b}_4$ and center weights z_1, z_3 are *curvature continuous* if

$$\frac{A_1}{z_1^2 \ell_1^3} = \frac{A_3}{z_3^2 \ell_2^3},$$ (6.10)

where $A_i = \text{area}(\mathbf{b}_{i-1}, \mathbf{b}_i, \mathbf{b}_{i+1})$ and $\ell_i = \text{length}(\mathbf{b}_i, \mathbf{b}_{i+1})$.

Note that (6.10) implies that the composite curve must be *convex*, since the signs of A_1 and of A_3 must be the same! We also use the term G^2 continuity, or second order geometric continuity to describe this type of smoothness.

If we multiply the central weights z_1 and z_3 of our composite conic by the same factor, we will maintain curvature continuity. We may thus pull the piecewise curve closer to the control polygon, or push it away from it, still maintaining curvature continuity.

An interesting feature[6] of conics is that the *ratio* of $\kappa(0)$ and $\kappa(1)$ is independent of the weight z_1:

$$\frac{\kappa(0)}{\kappa(1)} = \left[\frac{\ell_1}{\ell_0}\right]^3.$$ (6.11)

---

[6]This property was pointed out to me by a referee who suggested improvements on the paper [52].

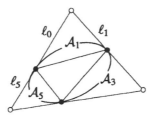

Figure 6.7. Closed conic splines: curvature continuity imposes additional conditions on the curve's geometry.

This has an implication on *closed* conic splines: consider the closed curve shown in Fig. 6.7. Curvature continuity demands that the following three conditions be met:

$$\frac{\mathcal{A}_1}{z_1^2 \ell_1^3} = \frac{\mathcal{A}_3}{z_3^2 \ell_2^3},$$

$$\frac{\mathcal{A}_3}{z_3^2 \ell_3^3} = \frac{\mathcal{A}_5}{z_5^2 \ell_4^3},$$

$$\frac{\mathcal{A}_5}{z_5^2 \ell_5^3} = \frac{\mathcal{A}_1}{z_1^2 \ell_0^3}.$$

If we multiply all left hand sides and all right hand sides, we obtain:

$$\ell_1 \ell_3 \ell_5 = \ell_2 \ell_4 \ell_0.$$

This condition — with an obvious extension for curves with more than three segments — limits the geometry of closed conic spline curves with curvature continuity. If only the $\mathbf{b}_{2i+1}$ were given, and we would be asked for permissible locations for the $\mathbf{b}_{2i}$, one possible choice for them is to set

$$\frac{\ell_{2i-1}}{\ell_{2i}} = \frac{\mathcal{A}_{2i-1}}{\mathcal{A}_{2i+1}},$$

where $\mathcal{A}_{2i+1}$ is the area of the triangle formed by $\mathbf{b}_{2i}, \mathbf{b}_{2i+1}, \mathbf{b}_{2i+2}$.

6.6 Interpolation with G^2 Conic Splines

We now consider an interpolation problem where the given data are:

- two points $\mathbf{b}_0$ and $\mathbf{b}_4$,
- a tangent at $\mathbf{b}_0$ and $\mathbf{b}_4$,

Figure 6.8. Conic interpolants: two conics may be used to interpolate to given positions, tangents, and curvatures.

- the intersection **d** of the two tangents,
- curvatures κ_0 and κ_2 at $\mathbf{b}_0$ and $\mathbf{b}_4$,
- another tangent **T**.

This configuration is assumed to be convex. Can we find a conic that interpolates to this data set? In general, we cannot — the given information is too much for one conic to cope with.

But we may use *two* conics: this approach is due to H. Pottmann [123]. He constructs two G^2 conics in standard form that join at a point $\mathbf{b}_2$, see Fig. 6.8.

With the notation from that figure, the condition for curvature continuity becomes

$$\frac{\mathcal{A}_1}{z_1^2 \ell_1^3} = \frac{\mathcal{A}_3}{z_3^2 \ell_2^3},$$

from which we conclude that

$$\frac{\ell_0^3}{\ell_3^3} \frac{\kappa_0}{\kappa_2} = \frac{\ell_1^3}{\ell_2^3}.$$

This allows us to compute ℓ_1/ℓ_2, and we have thus found the point $\mathbf{b}_2$. The missing weights z_1 and z_3 are found from the interpolatory conditions

$$\kappa_0 = \frac{\mathcal{A}_1}{z_1^2 \ell_0^2} \quad \text{and} \quad \kappa_2 = \frac{\mathcal{A}_3}{z_3^2 \ell_3^2}.$$

The tangent **T** might not be given in a particular application, and Pottmann [123] points out several ways to automate its choice.

When is it possible to solve our initial problem with *one* conic? For that to be possible, the prescribed data cannot be independent; rather, they must satisfy the restriction

$$\frac{\kappa_0}{\kappa_2} = \frac{||\mathbf{d} - \mathbf{b}_0||^3}{||\mathbf{d} - \mathbf{b}_4||^3},$$

as follows from (6.11). The number

$$I = \frac{\kappa_2}{\kappa_0} \frac{||\mathbf{d} - \mathbf{b}_0||^3}{||\mathbf{d} - \mathbf{b}_4||^3} \qquad (6.12)$$

thus measures the "non-conicness" of the given data. If it equals one, they permit an interpolating conic.

Pottmann [123] made the interesting geometric observation that the number I is invariant under projective transformations. In particular, the property $I = 1$ — indicating curvature continuity — is a projective invariant, a fact that is known as the Memke-Smith theorem. This theorem is somewhat surprising since individual curvatures are certainly not projectively invariant as they involve length and area computations.

Another class of interpolating piecewise conics are the so-called *biarcs*. Instead of using two general conics as in Section 6.6, we can specify that they should be *circles*. A biarc thus consists of two circular arcs and solves the following interpolation problem: given two points $\mathbf{b}_0$ and $\mathbf{b}_4$, tangents at $\mathbf{b}_0$ and $\mathbf{b}_4$, find two circular arcs which interpolate to the given data. They will join at a point $\mathbf{b}_2$ such that the two arcs are G^1 there. The point $\mathbf{b}_2$ is not part of the interpolation problem; once it has been determined, the interpolation problem is solved. Of course, biarcs are only G^1 — if two circular arcs were G^2, they would be *one* circle!

Some analysis of the problem reveals that all possible choices for $\mathbf{b}_2$ are restricted to lie on a circle through $\mathbf{b}_0$ and $\mathbf{b}_4$. Several choices for $\mathbf{b}_2$ are meaningful. Since the two arcs will differ in their curvatures, it is natural to fix $\mathbf{b}_2$ such that the difference in the "left" and "right" curvatures is minimized. This results in $\ell_1 = \ell_2$, using the notation from Fig. 6.8. Literature on biarcs, including 3D ones: [28], [89], [111], [106], [113], [112], [141].

While biarc curves are not G^2, they still are important in practical applications such as numerically controlled tools: those are able to move along straight lines and circular arcs.

6.7 Higher Order Smoothness

We have seen how to join conics so as to achieve G^1 or G^2 continuity. There are more ways of doing this; we report an approach that is due to Geise and Juettler [79].

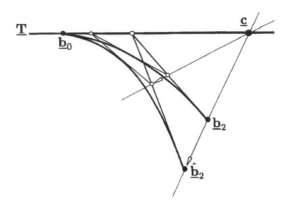

Figure 6.9. G^2 continuity: the two shown conics are related by a central collineation.

Referring to Fig. 6.9, we see two conics (in projective space) that share a common tangent $\underline{\mathbf{T}}$. Pick a point $\underline{\mathbf{c}}$ on this tangent, and define a *central collineation* (see Fig. 2.7) with axis $\underline{\mathbf{T}}$ and center $\underline{\mathbf{c}}$. Then the shown two conics are G^2 continuous in the following sense: in a "small" neighborhood of $\underline{\mathbf{b}}_0$, pick three points on each conic. If the neighborhood is chosen small enough, the three points will coincide. This is called "contact of order two", and is the most general definition of G^2 continuity. In the context of

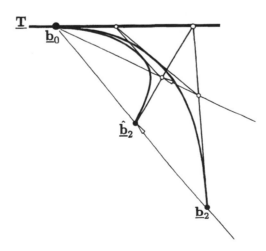

Figure 6.10. G^3 continuity: the center of the central collineation is at $\underline{\mathbf{b}}_0$.

affine parabolas, a commonly used G^2 condition is to choose **c** at infinity; then the central collineation becomes a shear.

An increase in the smoothness between two conics may be achieved as follows: referring to Fig. 6.10, we now choose a central collineation with center $\underline{\mathbf{b}}_0$. This ensures that the conics have G^3 continuity, or contact of order three.

Both kinds of geometric continuity can be used to refer to composite conics: just take a G^2 or G^3 piecewise conic and change one of its segments according to the above constructions. The resulting piecewise curve will still be G^2 or G^3.

6.8 Problems

1. Verify (6.9) for the case of a quarter circle in rational Bézier form. Then prove the general statement.

2. Devise a different system for estimating the $\mathbf{b}_{2i}$ from given $\mathbf{b}_{2i\pm1}$ in Section 6.5, using the idea of G^2 conic splines.

3. Numerically, (6.12) will rarely yield $I = 1$. Discuss what tolerances might be employed in determining if *one* conic can solve the G^2 interpolation problem.

7

Rational Bézier curves

Bézier curves may be viewed as the backbone of all piecewise polynomial curve schemes; their rational counterparts enjoy a similar role for piecewise rational curves. Every piecewise polynomial or rational curve may be broken down into a collection of Bézier curves, and thus their study is crucial to the understanding of the whole field of NURBS.

Historically, Bézier curves — in their polynomial, or integral, version — are due to P. de Casteljau. In 1957, he invented what is now known as the de Casteljau algorithm, laid down in an internal report of the Citroën automotive company. His algorithm relies on the concept of constant ratios, and is thus tied in closely with affine geometry. Rational Bézier curves, however, may be evaluated using the concept of cross ratios, the fundamental invariant of projective geometry.[1]

7.1 The Bernstein Form

Any n–th degree polynomial curve $\underline{b}(t)$ in $I\!\!P^3$ may be expressed in Bernstein form:

$$\underline{b}(t) = \sum_{i=0}^{n} \underline{b}_i B_i^n(t); \quad \underline{b}_i \in I\!\!P^3, \tag{7.1}$$

[1]But when I sent de Casteljau a preprint of [50], he was not too enthusiastic about the new version of his algorithm: he just did not like the idea of using cross ratios...

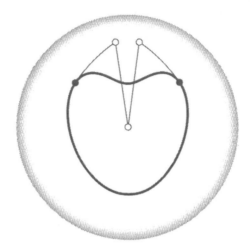

Figure 7.1. Bézier curves: a degree four (quartic) example.

where the B_i^n are the *Bernstein polynomials*

$$B_i^n(t) = \binom{n}{i} t^i (1-t)^{n-i}; \quad i = 0, \ldots, n. \tag{7.2}$$

It is customary to add the following to this definition:

$$B_i^n(t) \equiv 0 \quad \text{if} \quad i \notin \{0, ldots, n\}.$$

We note two important properties of Bernstein polynomials: they form a *partition of unity*:

$$\sum_{i=0}^{n} B_i^n(t) \equiv 1 \tag{7.3}$$

and they satisfy the recursion

$$B_i^n(t) = (1-t)B_{i-1}^{n-1}(t) + tB_i^{n-1}(t); \quad B_{-1}^n(t) = B_{n+1}^n(t) \equiv 0. \tag{7.4}$$

for more properties, see [57].

Figures 7.1, 7.2, and 7.3 give some examples of Bézier curves in $I\!P^2$; the 3D cases have a similar flavor. We show the complete curves — in most practical applications, one would only be interested in the segments "inside" the control polygon, i.e., those points corresponding to $t \in [0, 1]$ in (7.1).

We may project (7.1) into affine three-space: we will then obtain a *rational Bézier curve* $\mathbf{b}(t)$. This follows the same development as that of

Figure 7.2. Bézier curves: another quartic example.

Figure 7.3. Bézier curves: a quintic example.

rational quadratics in Section 5.1, and yields

$$b(t) = \frac{\sum_{i=0}^{n} w_i b_i B_i^n(t)}{\sum_{i=0}^{n} w_i B_i^n(t)}; \quad b_i \in I\!\!E^3. \tag{7.5}$$

The w_i are now called *weights*. If they are all positive, the curve lies in the convex hull of the control polygon $b_0, \ldots, b_n$.[2] This is called the *convex hull property* of Bézier curves.

If one weight, w_k, say, is increased, the curve is "pulled" towards b_k.

We can make a more precise statement as follows: let $b(t)$ be a point on some rational Bézier curve, and let $\bar{b}(t)$ be a point on a curve with the same polygon and set of weights, except that w_k is replaced by cw_k. Then the three points $b(t), \bar{b}(t)$, and b_k are collinear. This is easily seen in a projective context: then we have to show that $\det[\underline{b}(t), \underline{\bar{b}}(t), \underline{b}_k] = 0$. Writing this out in detail, we have

$$
\begin{aligned}
\det[\underline{b}(t), \quad \underline{\bar{b}}(t), \quad \underline{b}_k] \;=\; & \det \left[\sum_{\substack{i=0 \\ i \neq k}}^{n} \underline{b}_i B_i^n(t) + \underline{b}_k B_k^n(t), \right. \\
& \qquad \left. \sum_{\substack{i=0 \\ i \neq k}}^{n} \underline{b}_i B_i^n(t) + c\underline{b}_k B_k^n(t), \quad \underline{b}_k \right] \\
=\; & \det[\underline{b}_k(1-c)B_k^n(t), \quad \underline{\bar{b}}(t), \quad \underline{b}_k] \\
=\; & 0,
\end{aligned}
$$

since the first and the last columns of the last determinant are multiples of each other.[3]

Collinearity is preserved by going from projective to affine space, and so we have that $\bar{b}(t)$ is a barycentric combination of $b(t)$ and b_k. As w_k increases, $\bar{b}(t)$ is "pulled towards" b_k, whereas for decreasing w_k, its is "pushed away" from b_k.

We thus have two different ways to design rational Bézier curves: we may move a control point $\underline{b}_k$, or we may change a weight w_k. A third way is given by the introduction of *weight points*

$$q_i = \frac{(w_i b_i + w_{i+1} b_{i+1})}{w_i + w_{i+1}}. \tag{7.6}$$

---

[2]The convex hull of the polygon is defined as the set of all points p that can be written as $p = \alpha_0 b_0 + \ldots + \alpha_n b_n$ with $\alpha_0 + \ldots + \alpha_n = 1$ and $\alpha_i \geq 0$.

[3]A similar technique was used in Section 4.10.

Figure 7.4. Designing with rational Bézier curves: top: a nonrational curve of degree seven; middle: made rational. All weights equal one except $w_4 = w_5 = 0.1$. The weight points are shown as diamonds; bottom: two control points are moved (integral case).

While the weights determine the weight points, the converse is also true: given the weight points, we may compute a set of corresponding weights by observing that

$$\text{ratio}(\mathbf{b}_i, \mathbf{q}_i, \mathbf{b}_{i+1}) = \frac{w_{i+1}}{w_i};$$

refer to equation (2.21) for the definition of the ratio of three collinear points. Thus changing a weight point (and recomputing the weights) may be also used as a design tool.

Fig. 7.4 gives an impression of how rational Bézier curves may be used in a design situation. With nonrational curves, the standard interactive tool is the displacement of control points. This is still possible for rational

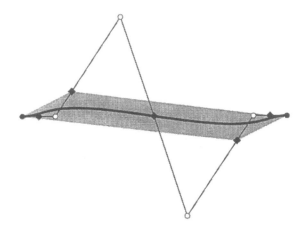

Figure 7.5. Weight points and convex hulls: if we use the weight points (diamonds), we may obtain a smaller size convex hull than by just using the control points.

curves, of course. Additional flexibility is gained by the ability to change the weights as well. This is in fact a different design tool: when a control point is moved, all curve points move parallel to its displacement vector. When a weight is changed, all curve points move towards the corresponding control point or away from it. It may be more intuitive for a designer to work with the weight points (geometric) than directly dealing with the weights (algebraic).

The weight points q_i should be located between b_i and b_{i+1}: if they are not, some weights will become negative.

The weight points may be used in the definition of a tighter convex hull for rational Bézier curves: in case of all weights w_i being positive, he curve lies in the convex hull of the point set $\{b_0, q_0, q_1, \ldots, q_{n-1}, b_n\}$. This follows from the fact that the q_i are intermediate points in the subdivision process for $t = \frac{1}{2}$, see [54] and Section 7.10. The convex hull of the set $\{b_0, q_0, q_1, \ldots, q_{n-1}, b_n\}$. is smaller than that of the control points. See Fig. 7.5 for an illustration. Smaller convex hulls may be advantageous for algorithms like clipping or curve/plane intersections.

If some of the w_i differ in sign, the denominator

$$w(t) = \sum_{i=0}^{n} w_i B_i^n(t)$$

of (7.5) may become zero, causing a singularity. Geometrically, this means that the projective curve $\underline{b}(t)$ crosses (or touches) the plane that is mapped

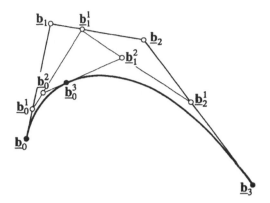

Figure 7.6. The projective de Casteljau algorithm: a point on a curve is constructed by repeated linear interpolation.

to the affine plane at infinity, much in the same way in which we discussed the asymptotes of hyperbolas, see Section 3.5. Keep in mind that then the control polygon will also cross the plane at infinity!

7.2   The de Casteljau algorithm

In projective space, $\underline{\mathbf{b}}(t)$ may be evaluated by a *projective de Casteljau algorithm*:

$$\underline{\mathbf{b}}_i^r(t) = (1-t)\underline{\mathbf{b}}_i^{r-1}(t) + t\underline{\mathbf{b}}_{i+1}^{r-1}(t); \begin{cases} r = 1, \ldots, n \\ i = 0, \ldots, n-r \end{cases} \tag{7.7}$$

with $\underline{\mathbf{b}}_i^0(t) = \underline{\mathbf{b}}_i$. Then $\underline{\mathbf{b}}_0^n(t) = \underline{\mathbf{b}}(t)$ is the point with parameter value t. Fig. 7.6 illustrates the cubic case $n = 3$.[4] This de Casteljau algorithm is also defined in $\mathbb{E}^3$; there it yields polynomial Bézier curves — for more details, see [57].

If we introduce

$$\underline{\mathbf{q}}_i = (\underline{\mathbf{b}}_i + \underline{\mathbf{b}}_{i+1})/2,$$

we observe that

$$\mathrm{cr}(\underline{\mathbf{b}}_i, \underline{\mathbf{b}}_i^1, \underline{\mathbf{q}}_i, \underline{\mathbf{b}}_{i+1}) = \frac{t}{1-t}, \quad \text{all } i.$$

---

[4]The careful observer may notice that the collinear $\underline{\mathbf{b}}_i^r, \underline{\mathbf{b}}_i^{r+1}, \underline{\mathbf{b}}_{i+1}^r$ are not in the same ratio for all r: this is intentional, as ratios are not a projective concept.

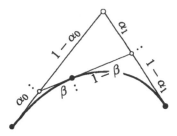

Figure 7.7. The rational de Casteljau algorithm: some of the involved barycentric combinations.

Just as we did to prove (4.6), we can define

$$\underline{\mathbf{q}}_i^r = (\underline{\mathbf{b}}_i^r \wedge \underline{\mathbf{b}}_{i+1}^r) \wedge (\underline{\mathbf{q}}_i^{r-1} \wedge \underline{\mathbf{q}}_{i+1}^{r-1}) \tag{7.8}$$

and obtain that

$$\underline{\mathbf{q}}_i^r = \frac{1}{2}\underline{\mathbf{b}}_i^r + \frac{1}{2}\underline{\mathbf{b}}_{i+1}^r. \tag{7.9}$$

As a consequence,

$$\mathrm{cr}(\underline{\mathbf{b}}_i^r, \underline{\mathbf{b}}_i^{r+1}, \underline{\mathbf{q}}_i^r, \underline{\mathbf{b}}_{i+1}^r) = \frac{t}{1-t}. \tag{7.10}$$

This shows that the projective de Casteljau algorithm (7.7) is projectively invariant: the point $\underline{\mathbf{b}}_0^n(t)$ can be found by intersections and cross ratio constructions only. The points $\underline{\mathbf{q}}_i$ are, of course, the projective pre-images of the weight points $\mathbf{q}_i$.

We could also project every step of the projective de Casteljau algorithm into $I\!\!E^3$ and obtain the *rational de Casteljau algorithm*:

$$\mathbf{b}_i^r(t) = \frac{(1-t)w_i^{r-1}\mathbf{b}_i^{r-1}(t) + tw_{i+1}^{r-1}\mathbf{b}_{i+1}^{r-1}(t)}{w_i^r(t)} \qquad \left\{ \begin{array}{l} r = 1,\dots,n \\ i = 0,\dots,n-r \end{array} \right. \tag{7.11}$$

with $\mathbf{b}_i^0(t) = \mathbf{b}_i$ and $w_i^r(t) = (1-t)w_i^{r-1}(t) + tw_{i+1}^{r-1}(t)$. Then $\mathbf{b}_0^n(t) = \mathbf{b}(t)$ is the point with parameter value t. Here, care must be taken in the case of t being outside the interval $[0, 1]$, as some of the intermediate values may be vectors![5] Example 7.1 illustrates this: the projective point $\underline{\mathbf{b}}_2^1$ projects to an affine vector, and thus would need special case handling. It is best, therefore, to carry out all computations in projective space.

---

[5]The intermediate points could also be *close* to being vectors, i.e., having a very small third component. This may cause numerical problems.

The following is a de Casteljau scheme for the control polygon

$$\begin{bmatrix} -1 \\ 1 \end{bmatrix}, \begin{bmatrix} -1 \\ 0 \end{bmatrix}, \begin{bmatrix} 1 \\ 0 \end{bmatrix}, \begin{bmatrix} 1 \\ 1 \end{bmatrix},$$

weights $1, 2, 2, 1$ and $t = 2$. The computation is carried out in the projective plane.

$$\begin{bmatrix} -1 \\ 1 \\ 1 \end{bmatrix}$$

$$\begin{bmatrix} -2 \\ 0 \\ 2 \end{bmatrix} \quad \begin{bmatrix} -3 \\ -1 \\ 3 \end{bmatrix}$$

$$\begin{bmatrix} 2 \\ 0 \\ 2 \end{bmatrix} \quad \begin{bmatrix} 6 \\ 0 \\ 2 \end{bmatrix} \quad \begin{bmatrix} 15 \\ 1 \\ 1 \end{bmatrix}$$

$$\begin{bmatrix} 1 \\ 1 \\ 1 \end{bmatrix} \quad \begin{bmatrix} 0 \\ 2 \\ 0 \end{bmatrix} \quad \begin{bmatrix} -6 \\ 4 \\ -2 \end{bmatrix} \quad \begin{bmatrix} -27 \\ 7 \\ -5 \end{bmatrix}$$

The last entry projects to the affine point $\mathbf{b}(2) = [27/5, -7/5]^T$. Note that $\underline{\mathbf{b}_2^1}$'s third coordinate does not pose a problem at all!

Example 7.1. An example of the de Casteljau algorithm.

Finally, we present a different version of the rational de Casteljau algorithm described by Boehm [23]. It is sufficient to explain the basic idea for the case $n = 2$. There, we have (cf. Fig. 7.7):

$$\mathbf{b}_0^2 = \frac{(1-t)w_0^1}{w_0^2}\mathbf{b}_0^1 + \frac{tw_1^1}{w_0^2}\mathbf{b}_1^1 =: (1-\beta)\mathbf{b}_0^1 + \beta\mathbf{b}_1^1,$$

$$\mathbf{b}_0^1 = \frac{(1-t)w_0}{w_0^1}\mathbf{b}_0 + \frac{tw_1}{w_0^1}\mathbf{b}_1 =: (1-\alpha_0)\mathbf{b}_0^1 + \alpha_0\mathbf{b}_1^1,$$

$$\mathbf{b}_1^1 = \frac{(1-t)w_1}{w_1^1}\mathbf{b}_1 + \frac{tw_2}{w_1^1}\mathbf{b}_2 =: (1-\alpha_1)\mathbf{b}_1 + \alpha_1\mathbf{b}_2.$$

It is straightforward to verify that

$$\frac{\beta}{1-\beta} = \frac{\alpha_0}{1-\alpha_1}. \tag{7.12}$$

This observation may be used to reformulate the rational de Casteljau algorithm: once we have computed the $\alpha_i; i = 0, \ldots, n-1$, we may continue

with barycentric combinations that use only these terms.[6] The following
scheme shows, for $n = 4$, the involved ratios.

$$
\begin{array}{llll}
\alpha_0/(1 - \alpha_0) & & & \\
\alpha_1/(1 - \alpha_1) & \alpha_0/(1 - \alpha_1) & & \\
\alpha_2/(1 - \alpha_2) & \alpha_1/(1 - \alpha_2) & \alpha_0/(1 - \alpha_2) & \\
\alpha_3/(1 - \alpha_3) & \alpha_2/(1 - \alpha_3) & \alpha_1/(1 - \alpha_3) & \alpha_0/(1 - \alpha_3).
\end{array}
\tag{7.13}
$$

In case of all $w_i = 1$, all these ratios become the familiar $t/(1 - t)$ from the
integral de Casteljau algorithm.

7.3   Degree Elevation

A projective Bézier curve $\underline{\mathbf{b}}^n(t) = \sum_{i=0}^{n} \underline{\mathbf{b}}_i B_i^n(t)$ of degree n may be expressed as a Bézier curve $\sum_{i=0}^{n+1} \underline{\mathbf{b}}_i^* B_i^{n+1}$ of degree $n+1$. The control points
$\underline{\mathbf{b}}_i^*$ are given by

$$
\underline{\mathbf{b}}_i^* = \frac{i}{n+1}\underline{\mathbf{b}}_{i-1} + (1 - \frac{i}{n+1})\underline{\mathbf{b}}_i; \quad i = 0, \ldots, n+1. \tag{7.14}
$$

This is seen by multiplying $\sum_{i=0}^{n} \underline{\mathbf{b}}_i B_i^n(t)$ by $[t + (1 - t)]$ and comparing
powers of t^i and $(1 - t)^{n+1-i}$.

We can project this into affine space, and obtain the *degree elevation*
formulas for rational Bézier curves:

$$
\mathbf{b}_i^* = \frac{iw_{i-1}\mathbf{b}_{i-1} + (n + 1 - i)w_i\mathbf{b}_i}{iw_{i-1} + (n + 1 - i)w_i}; \quad i = 0, \ldots, n+1, \tag{7.15}
$$

$$
w_i = iw_{i-1} + (n + 1 - i)w_i. \tag{7.16}
$$

Fig. 7.8 gives an example of a conic that is degree elevated to rational
cubic form.

We may degree elevate a rational Bézier curve repeatedly, resulting in a
sequence of weights and control polygons. Provided that all weights are of
the same sign, the control polygons will converge to the curve. This follows
from the corresponding results for polynomial curves, see [49] or [57].

We proved (7.14) by multiplying the equation of a Bézier curve by the
"harmless" constant 1, written as $1 = [t + (1 - t)]$. But we may multiply
$\underline{\mathbf{b}}(t)$ by any other nonzero function $\rho(t)$, and would not change its shape!
The function ρ may be polynomial, thus raising the degree of the original
curve. If ρ has zeroes, we will introduce base points, see Section 7.7.

---

[6]We should note that this reformulation is more of a geometric than of a computa-
tional nature: the algorithm proceeds as before!

Figure 7.8. Rational degree elevation: a rational quadratic is degree elevated to rational cubic.

7.4 Degree Reduction

In general, we will not be able to lower the degree of a given rational Bézier curve. For example, a cubic, which might have an inflection point, cannot be written as a rational quadratic, because a conic cannot have an inflection point. The process of degree reduction will therefore be approximative. For the special case of approximating a rational cubic by a rational quadratic, see section 8.1.

We will choose to perform degree reduction in projective space — there, we only have to deal with polynomials. Our aim is therefore the following: given a Bézier curve $\underline{b}(t)$ with control vertices $\underline{b}_i$; $i = 0, \ldots, n$, find a Bézier curve $\hat{\underline{b}}(t)$ with control vertices $\hat{\underline{b}}_i$; $i = 0, \ldots, n-1$ that approximates the first curve. Several methods exist for this purpose, although they have all been developed in a Euclidean context. They can all be cast in the following setting.

Let us pretend that the $\underline{b}_i$ were obtained from the $\hat{\underline{b}}_i$ by the process of degree elevation (this is not true, in general, but makes a good working assumption). Then they would be related by

$$\underline{b}_i = \frac{i}{n}\hat{\underline{b}}_{i-1} + \frac{n-i}{n}\hat{\underline{b}}_i; \quad i = 0, 1, \ldots, n. \tag{7.17}$$

This equation can be used to derive two recursive formulas for the generation of the $\hat{\underline{b}}_i$ from the $\underline{b}_i$:

$$\hat{\underline{b}}_i^{\text{left}} = \frac{n\underline{b}_i - i\hat{\underline{b}}_{i-1}^{\text{left}}}{n-i}; \quad i = 0, 1, \ldots, n-1 \tag{7.18}$$

and

$$\hat{\underline{b}}_{i-1}^{\text{right}} = \frac{n\underline{b}_i - (n-i)\hat{\underline{b}}_i^{\text{right}}}{i}; \quad i = n, n-1, \ldots, 1. \tag{7.19}$$

The $\hat{\underline{\mathbf{b}}}_i^{\text{left}}$ perform satisfactorily near $\underline{\mathbf{b}}_0$, whereas the $\hat{\underline{\mathbf{b}}}_{i-1}^{\text{right}}$ do well near $\underline{\mathbf{b}}_n$. It seems a good idea therefore, to *blend* the two sets in order to obtain a final control net $\hat{\underline{\mathbf{b}}}_i$:

$$\hat{\underline{\mathbf{b}}}_i = (1 - \alpha_i)\hat{\underline{\mathbf{b}}}_i^{\text{left}} + \alpha_i\hat{\underline{\mathbf{b}}}_i^{\text{right}}. \tag{7.20}$$

R. Forrest [76] has suggested to set $\alpha_i = 0$ for $i < n/2$ and $\alpha_i = 1$ for $i > n/2$, with a case distinction for $\alpha_{n/2}$ when n is odd. Another choice with reasonable outcome is $\alpha_i = i/(n-1)$. An "optimal" choice is given by

$$\alpha_i = 2^{1-2n}\sum_{j=0}^{i}\binom{2n}{2j}; \quad i = 0,\ldots,n-1. \tag{7.21}$$

In Euclidean space, this choice ensures that we obtain the best approximation (in the Chebychev sense) of degree $n - 1$ to the given polynomial of degree n, see Eck[47]. In projective space, such claims are meaningless, but we should still expect (7.21) to perform well.

The first appearance of degree reduction is in Forrest [76]. Eck's geometric treatment is similar to more algebraic methods by Watkins and Worsey [159] and Lachance [98]. See also [30] and [48].

7.5   Reparametrization

We have seen that an arc of a conic does not have a unique parametric representation: depending on the choice of the shoulder tangent, each weight w_i could be multiplied by the power c^i of a nonzero constant c: this would not change the geometry of the curve, but it would change how the parameter traverses it. See Section 4.9 for details. The same principle also serves to reparametrize Bézier curves, in projective space as well after being projected into affine space.

We can write a projective Bézier curve in the following two ways:

$$\sum_{i=0}^{n}\underline{\mathbf{b}}_i B_i^n(t) = \sum_{i=0}^{n}c^i\underline{\mathbf{b}}_i B_i^n(\hat{t}),$$

which means that a rational Bézier curve can be written as

$$\frac{\sum_{i=0}^{n} w_i\mathbf{b}_i B_i^n(t)}{\sum_{i=0}^{n} w_i B_i^n(t)} = \frac{\sum_{i=0}^{n} c^i w_i\mathbf{b}_i B_i^n(\hat{t})}{\sum_{i=0}^{n} c^i w_i B_i^n(\hat{t})}.$$

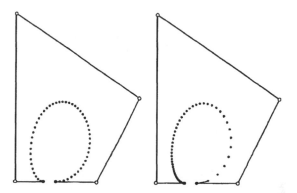

Figure 7.9. Reparametrization: a rational Bézier curve is sampled at 50 equal parameter intervals. Left, $c = 1$; right: c=1/3.

Reparametrization of a rational Bézier curve changes the rate in which the parameter t traverses the curve; Fig. 7.9 illustrates.

After a curve has been reparametrized, it will have new weight points $\hat{\mathbf{q}}_i$. Just as in the conic case, the old and new weight points are related by a constant cross ratio relation:

$$\mathrm{cr}(\mathbf{b}_i, \mathbf{q}_i, \hat{\mathbf{q}}_i, \mathbf{b}_{i+1}) = \frac{1}{c}. \tag{7.22}$$

The old and new parameters of the curve are related by the same rational linear, or Moebius, transformation (4.37) that was used for the conic case:

$$\hat{t} = \frac{ct}{1 + (c - 1)t}. \tag{7.23}$$

A graph of the rational linear function (7.23) is shown in Fig. 7.10.[7]

---

[7]It is interesting to note that another, very similar class, of reparametrizations is used in grid generation for finite element analysis: there, one encounters transformations of the form

$$\hat{t} = \frac{e^{kt} - 1}{e^k - 1}. \tag{7.24}$$

The graphs of these functions, for varying k, looks much the same as those in 7.10. For our purposes, it is important to note that (7.23) produces yet another rational curve, whereas (7.24) takes us out of that class.

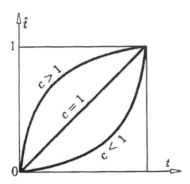

Figure 7.10. Reparametrization: the relationship between old and new parameter depending on the constant c.

A rational Bézier curve with positive weights may be reparametrized so that the end weights are unity: this is achieved by selecting

$$c = \sqrt[n]{\frac{w_n}{w_0}}.$$

Curves in this form are said to be in *standard form*.

Reparametrization and degree elevation together can cause an effect that cannot be found in polynomial curves. Take a rational cubic, for example, and degree elevate it. We could take that same rational cubic,

Figure 7.11. Rational ambiguities: a rational quadratic may be written as two different rational cubics using degree elevation and reparametrization.

reparametrize it and then degree elevate it. In this way, we obtain two different rational quartic representations of the same curve! See Fig. 7.11 for an illustration. The polynomial case is quite different: two polynomial Bézier curves of the same degree are equal if and only if their control polygons are the same.

7.6 Nonparametric Curves

A functional, or nonparametric, integral Bézier curve is defined by

$$y = \sum_{i=0}^{n} b_i B_i^n(x).$$

It is easily represented in terms of a parametric Bézier curve: its Bézier points are given by $\mathbf{b}_i = [i/n, b_i]^T$. This works since the x-coordinates satisfy

$$x = \sum_{i=0}^{n} \frac{i}{n} B_i^n(x). \tag{7.25}$$

The situation in the rational case is somewhat more complicated. Let

$$y = \frac{\sum w_i b_i B_i^n(x)}{\sum w_i B_i^n(x)} \tag{7.26}$$

be a rational function of x. Can we write it as a parametric rational Bézier curve, and if so, what are its Bézier points $\mathbf{b}_i$ and the corresponding weights v_i? Written in parametric form, (7.26) takes the form

$$\left[\begin{array}{c} x(u) \\ y(u) \end{array} \right] = \left[\begin{array}{c} u \\ \dfrac{\sum w_i b_i B_i^n(u)}{\sum w_i B_i^n(u)} \end{array} \right] = \left[\begin{array}{c} \dfrac{u \sum w_i B_i^n(u)}{\sum w_i B_i^n(u)} \\ \dfrac{\sum w_i b_i B_i^n(u)}{\sum w_i B_i^n(u)} \end{array} \right]$$

Using the identity $u B_i^n(u) = \frac{i+1}{n+1} B_{i+1}^{n+1}(u)$ and degree elevation, we obtain

$$\left[\begin{array}{c} x(u) \\ y(u) \end{array} \right] = \frac{\sum_{i=0}^{n+1} v_i \mathbf{b}_i B_i^{n+1}(u)}{\sum_{i=0}^{n+1} v_i B_i^{n+1}(u)},$$

Using degree elevation, the function $y = 1/(1 + x)$ may be written as

$$y = \frac{1 \cdot 1 \cdot B_0^1(x) + 2 \cdot \frac{1}{2} \cdot B_1^1(x)}{1 \cdot B_0^1(x) + 2 \cdot B_1^1(x)}.$$

We may write the segment over $0 \leq x \leq 1$ as a rational quadratic; its Bézier points are

$$\mathbf{b}_i = \begin{bmatrix} 0 \\ 1 \end{bmatrix}, \begin{bmatrix} 1/3 \\ 2/3 \end{bmatrix}, \begin{bmatrix} 1 \\ 1/2 \end{bmatrix},$$

and the corresponding weights are $v_i = 2, 3, 4$. We may standardize this curve and obtain weights $\hat{v}_i = 1, \frac{3}{4}\sqrt{2}, 1$. The middle weight is larger than one, indicating that our example function represents a hyperbola.

Example 7.2. A parametric representation of a rational linear function.

where

$$\mathbf{b}_i = \frac{1}{v_i} \begin{bmatrix} iw_{i-1} \\ iw_{i-1}b_{i-1} + (n+1-i)w_i b_i \end{bmatrix}; \quad i = 0, \ldots, n+1$$

and

$$v_i = iw_{i-1} + (n+1-i)w_i; \quad i = 0, \ldots, n+1.$$

In the integral case, the abscissae i/n of the Bézier points of a functional Bézier curve do not depend on the function; now they depend on the weights of the function under consideration. Thus a rational function $y = a(x)/b(x)$ with both a and b polynomials of degree n has a rational parametric representation of degree $n + 1$.

The abscissae iw_{i-1}/v_i are in the range $[0, 1]$. But note that they do not necessarily have to be increasing! For example, if the original weight sequence $\{w_i\}$ is $1, 10, 1, 10$, then the abscissae sequence $\{iw_{i-1}/v_i\}$ is $0, 1/31, 10/11, 3/13, 1$. See Example 7.2 and Problems for more cases.

7.7 Base Points

What happens if all components of a projective Bézier curve vanish simultaneously, yielding an expression of the form $\mathbf{b}(t) = \mathbf{0}$, clearly meaningless? In this situation, the corresponding parameter value t is called a *base point*. Thus a base point is a parameter value that does not map to a unique point in the range.

Let us study the case $t = 0$ instead of the general case. This is no restriction since we may employ subdivision (see Section 7.10) to make any point on a curve correspond to $t = 0$. If we have a base point for $t = 0$, then it follows that $\underline{\mathbf{b}} = \underline{\mathbf{0}}$ and our Bézier curve reduces to

$$
\begin{aligned}
\underline{\mathbf{b}}(t) &= \sum_{i=1}^{n} \underline{\mathbf{b}}_i \binom{n}{i} t^i (1-t)^{n-i} \\
&= t \sum_{i=1}^{n} \underline{\mathbf{b}}_i \binom{n}{i} t^{i-1} (1-t)^{n-i} \\
&\doteq \sum_{i=0}^{n-1} \underline{\mathbf{b}}_{i+1} \binom{n-1}{i+1} B_i^{n-1}(t).
\end{aligned}
$$

This has to be interpreted as follows: the base point reduced the degree of the curve by one. For $t = 0$, the curve now starts at $\underline{\mathbf{b}}_1$ instead of $\underline{\mathbf{b}}_0$.

For any curve in projective space, base points may be introduced by multiplying by an arbitrary function $\phi(t)$ — recall that this does not change the shape of the curve. So replacing $\underline{\mathbf{b}}(t)$ by $\phi(t)\underline{\mathbf{b}}(t)$ results in base points at every zero of $\phi(t)$. If ϕ is a rational polynomial, then so is $\phi(t)\underline{\mathbf{b}}(t)$, but now with base points present. Such polynomials are called *improperly parametrized*; Sederberg [135] discusses how to detect and to remove such parametrizations.

Another way of introducing a base point is by *degree elevation* . If we write a polynomial curve in projective space in monomial form: $\underline{\mathbf{x}}(t) = \underline{\mathbf{a}}_0 + \underline{\mathbf{a}}t + \ldots + \underline{\mathbf{a}}_n t^n$, then degree elevation amounts to formally adding a term in t^{n+1}, i.e., $\underline{\mathbf{x}}(t) = \underline{\mathbf{a}}_0 + \underline{\mathbf{a}}t + \ldots + \underline{\mathbf{a}}_n t^n + \underline{\mathbf{0}}t^{n+1}$. As we let t tend to infinity, the curve will tend to $\underline{\mathbf{0}}$, the coefficient of the highest power of t.[8] In this sense, degree elevation amounts to the introduction of a base point at $t = \infty$.

7.8 Derivatives

The derivative of a rational function seemingly involves the quotient rule; but we will avoid its use just as we did in the conic case. So we define **p** by

$$
\mathbf{p}(t) = w(t)\mathbf{b}(t), \tag{7.27}
$$

---

[8]This also follows from the fact that the nth derivative of an nth degree polynomial agrees with its value for $t = \infty$.

where $w(t)$ is the denominator of the rational Bézier curve $\mathbf{b}(t)$. Following the steps in Section 5.4, we obtain for the derivative $\dot{\mathbf{b}}(t)$:

$$\dot{\mathbf{b}}(t) = \frac{1}{w(t)}[\dot{\mathbf{p}}(t) - \dot{w}(t)\mathbf{b}(t)], \tag{7.28}$$

the dot denoting differentiation with respect to t. For the first derivatives at the endpoints of a rational Bézier curve, we find

$$\dot{\mathbf{b}}(0) = \frac{nw_1}{w_0}\Delta\mathbf{b}_0 \tag{7.29}$$

and

$$\dot{\mathbf{b}}(1) = \frac{nw_{n-1}}{w_n}\Delta\mathbf{b}_{n-1} \tag{7.30}$$

For higher derivatives, we differentiate (7.27) r times:

$$\mathbf{p}^{(r)}(t) = \sum_{j=0}^{r}\binom{r}{j}w^{(j)}(t)\mathbf{b}^{(r-j)}(t).$$

We can solve for $\mathbf{b}^{(r)}(t)$:

$$\mathbf{b}^{(r)}(t) = \frac{1}{w(t)}[\mathbf{p}^{(r)} - \sum_{j=1}^{r}\binom{r}{j}w^{(j)}(t)\mathbf{b}^{(r-j)}(t)]. \tag{7.31}$$

This is a recursive formula for the rth derivative of a rational Bézier curve. It only involves taking derivatives of polynomial curves, thus avoiding the quotient rule all together.

It is also possible to compute derivatives using the rational de Casteljau algorithm(7.11). We then obtain

$$\dot{\mathbf{b}}(t) = n\frac{w_0^{n-1}w_1^{n-1}}{[w_0^n]^2}[\mathbf{b}_1^{n-1} - \mathbf{b}_0^{n-1}]. \tag{7.32}$$

For a proof, see Floater [72]. Note that we do have to carry out the full rational de Casteljau algorithm; we could compute everything in projective space, up to $r = n - 1$, that is. Then we project into affine space so that we can apply (7.32).

For some algorithms, it is useful to know an upper bound for the derivative (the hodograph) of a curve. Sederberg and Wang [140] give the bound

$$\|\dot{\mathbf{b}}(t)\| \leq n\frac{\max\{w_i\}}{\min\{w_i\}}\max_{0\leq i,j\leq n}\{\|\mathbf{b}_i - \mathbf{b}_j\|\}, \quad 0 \leq t \leq 1. \tag{7.33}$$

Floater [72] gives

$$\|\dot{\mathbf{b}}(t)\| \leq n \left[\frac{\max\{w_i\}}{\min\{w_i\}} \right]^2 \max_{0 \leq i < n} \{\|\mathbf{b}_{i+1} - \mathbf{b}_i\|\} \qquad (7.34)$$

and notes that neither bound is sharper than the other. Of course, (7.34) is much faster to compute!

At the endpoints of a curve, derivatives are used to describe the curve's geometry. In terms of projective geometry, the *tangent* at $t = 0$ is given by the line through $\mathbf{b}_0$ and $\mathbf{b}_1$; the *osculating plane* at $t = 0$ is given by the plane $< \mathbf{b}_0, \mathbf{b}_1, \mathbf{b}_2 >$, with analogous results for $t = 1$. For the special case of a cubic, we see that the Bézier point $\mathbf{b}_1$ is the intersection of the osculating plane at $t = 1$ and the tangent at $t = 0$, and that $\mathbf{b}_2$ is the intersection of the osculating plane at $t = 0$ and the tangent at $t = 1$.

The concepts of tangents and osculating planes are projectively invariant; thus the same statements hold for the rational case.

7.9 Smoothness

We discussed C^1 continuity for piecewise conic curves in Section 6.1. Nothing new really happens when we move on to higher degree curves. Again, the concept of C^1 smoothness is more easily discussed in projective space as differentiability is preserved under projections into affine space.

7.10 Blossoms

We already introduced the concepts of blossoms for projective quadratic curves in Section 4.8. Quite analogously, we now define the blossom $\underline{\mathbf{b}}[t_1, \ldots, t_n]$ by applying the de Casteljau algorithm to the control polygon $\underline{\mathbf{b}}_0, \ldots, \underline{\mathbf{b}}_n$, but with a different argument t_i for each level of the algorithm. From the discussion in Section 4.8, it is clear that the value of $\underline{\mathbf{b}}[t_1, \ldots, t_n]$ does not depend on the order in which the t_i are fed into the algorithm. The blossom $\underline{\mathbf{b}}[t_1, \ldots, t_n]$ is thus a symmetric function of n variables. If all variables are equal: $t_1 = \ldots = t_n = t$, then the blossom agrees with the curve. It is also clear from its definition that the blossom is multilinear:

$$\underline{\mathbf{b}}[\ldots, (1 - \alpha)r + \alpha s, \ldots] = (1 - \alpha)\underline{\mathbf{b}}[\ldots, r, \ldots] + \alpha\underline{\mathbf{b}}[\ldots, s, \ldots]. \qquad (7.35)$$

What is not immediately clear, and in fact requires a thorough proof, is the fact that the blossom is uniquely associated with the given curve — it does not depend on its initial form, Bézier or other.

Figure 7.12. Blossoms: a cubic in $I\!P^2$ with two Bézier representations.

The original de Casteljau algorithm, with its intermediate points $\underline{\mathbf{b}}_i^r$, may be rewritten in blossom form; we give the cubic example:

$$
\begin{array}{llll}
\underline{\mathbf{b}}_0 & & & \\
\underline{\mathbf{b}}_1 & \underline{\mathbf{b}}_0^1 & & \\
\underline{\mathbf{b}}_2 & \underline{\mathbf{b}}_1^1 & \underline{\mathbf{b}}_0^2 & \\
\underline{\mathbf{b}}_3 & \underline{\mathbf{b}}_2^1 & \underline{\mathbf{b}}_1^2 & \underline{\mathbf{b}}_0^3
\end{array}
\;=\;
\begin{array}{llll}
\underline{\mathbf{b}}[0,0,0] & & & \\
\underline{\mathbf{b}}[0,0,1] & \underline{\mathbf{b}}[0,0,t] & & \\
\underline{\mathbf{b}}[0,1,1] & \underline{\mathbf{b}}[0,1,t] & \underline{\mathbf{b}}[0,t,t] & \\
\underline{\mathbf{b}}[1,1,1] & \underline{\mathbf{b}}[1,1,t] & \underline{\mathbf{b}}[1,t,t] & \underline{\mathbf{b}}[t,t,t].
\end{array}
$$

In general, the $\underline{\mathbf{b}}_i^r(t)$ are written in blossom form as

$$
\underline{\mathbf{b}}_i^r(t) = \underline{\mathbf{b}}[0^{<n-r-i>}, t^{<r>}, 1^{<i>}]
$$

The blossom can also be used to *subdivide* a Bézier curve: if we are interested in the control points $\underline{\mathbf{c}}_i$ of the curve segment corresponding to $[0, b]$, all we have to do is to evaluate the blossom:

$$
\underline{\mathbf{c}}_i = \underline{\mathbf{b}}[0^{<n-i>}, b^{<i>}]. \tag{7.36}
$$

($b = \frac{1}{2}$) was already presented for conics in Section 5.3.

If we are interested in the control points $\underline{\mathbf{c}}_i$ corresponding to an arbitrary interval $[a, b]$, we proceed in the same way:

$$
\underline{\mathbf{c}}_i = \underline{\mathbf{b}}[a^{<n-i>}, b^{<i>}]. \tag{7.37}
$$

Fig. 7.12 shows one cubic curve in $I\!P^2$ together with two different Bézier representations. Given one of them, the other one may be found by the use of (7.37).

Figure 7.13. Derivatives: a cubic together with its derivative — a gray curve on the horizon.

Blossoms may also be used to compute derivatives of Bézier curves in projective space. The following is the natural generalization of the conic case (4.32):

$$\frac{d^r}{dt^r}\underline{\mathbf{b}}(t) \,\hat{=}\, \underline{\mathbf{b}}[\infty^{<r>}, t^{<n-r>}]. \tag{7.38}$$

Fig. 7.13 shows a cubic in $I\!P^2$ with all z_i equal to one; its derivative is a quadratic which "lives" entirely on the horizon, or line at infinity.

An example with weights that differ from unity is shown in Fig. 7.14.

Special blossoms, called *osculants* may be defined at the curve's end-points:

$$\underline{\mathbf{o}}_l^r(t) = \underline{\mathbf{b}}[0^{<n-r>}, t^{<r>}], \quad \underline{\mathbf{o}}_r^r(t) = \underline{\mathbf{b}}[1^{<n-r>}, t^{<r>}]. \tag{7.39}$$

These are curves of degree r that agree in all r derivatives with those of the curve at $t = 0$ or $t = 1$, respectively. For $r = 1$, they are simply tangent lines. For $r = 2$, they are conics. The plane that is defined by such a conic osculant is called the *osculating plane* of the curve.

Fig. 7.15 shows two osculants $\underline{\mathbf{o}}_l^2$ and $\underline{\mathbf{o}}_r^2$ to a cubic.

Osculants may be defined for arbitrary parameter values; at parameter value s, we then have the osculant $\underline{\mathbf{b}}[s^{<n-r>}, t^{<r>}]$. For the special case of $s = \infty$, the osculants agree with derivatives.

The idea of osculants goes back to Jolles [93], see [26].

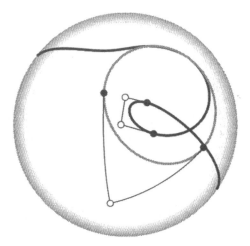

Figure 7.14. Derivatives: a cubic together with its derivative, shown in gray.

Figure 7.15. Osculants: the osculants at the endpoints of a cubic are shown.

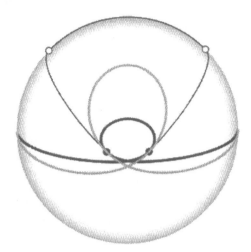

Figure 7.16. Infinite control points: two cubics with the same control polygons yet different factors for the points at infinity.

7.11   Control Vectors

In Section 5.5, we encountered control vectors (also known as infinite control points) as the limiting case of parallel tangents to a conic. The resulting rational curve representation contained both points and vectors. We can devise a similar form for rational Bézier curves (first suggested by K. Vesprille [151]). They will be of the form

$$\mathbf{b}(t) = \frac{\sum_{\text{points}} w_i \mathbf{b}_i B_i^n(t) + \sum_{\text{vectors}} \mathbf{v}_i B_i^n(t)}{\sum_{\text{points}} w_i B_i^n(t)}, \qquad (7.40)$$

see also [68]. The control vectors $\mathbf{v}_i$ do not have weights in this form; we may multiply each of them by a factor, and the curve will change accordingly. This is shown in Fig. 7.16: two cubics are shown with the same control polygon. Each has two points at infinity; the gray curve's infinite points were given a factor of two relative to the black curve's ones.

Note that at least one of the point weights w_i must be nonzero for (7.40) to be meaningful — otherwise a point (namely $\mathbf{b}(t)$) would have to equal a vector!

For theoretical purposes, the control vector form is attractive and has received some attention in the literature, see [121], [119], [120]. But they are not as useful in practical situations: as in the conic case, we have lost the convex hull property, and evaluation of (7.40) will require special case treatment. Also, control vectors cannot be incorporated into the current IGES standard, see Section 15.1.

7.12 Duality

We derived Bézier curves as a generalization of the line conic construction from Section 4.2. Of course, we could also generalize the point conic construction from Section 4.3. In the planar case, this leads to what was called *dual Bézier curves* by J. Hoschek [87].[9] Further literature: Boddoluri and Ravani [21] and [125].

To define these curves, we are given a set of *control lines* $\underline{\mathbf{B}}_i$ in the projective plane. We define a curve by defining its tangent $\underline{\mathbf{B}}(t)$ for every parameter value t:

$$\underline{\mathbf{B}}(t) = \sum_{i=0}^{n} \underline{\mathbf{B}}_i(t). \tag{7.41}$$

We say that (7.41) defines a *control line* Bézier curve. (In this context, we shall call the "usual" Bézier curves *control point* Bézier curves.) The curve is thus given as the envelope of its tangents. We see that the tangents at $t = 0$ and $t = 1$ are given by $\underline{\mathbf{B}}_0$ and $\underline{\mathbf{B}}_1$, respectively.

We can develop a de Casteljau algorithm for control line curves just as we could for control point curves. Any two subsequent lines $\underline{\mathbf{B}}_i$ and $\underline{\mathbf{B}}_{i+1}$ define a point $\underline{\mathbf{b}}_i = \underline{\mathbf{B}}_i \wedge \underline{\mathbf{B}}_{i+1}$, which in turn carries a pencil of lines through it. In each of these pencils, we can define a line $\underline{\mathbf{Q}}_i = \frac{1}{2}\underline{\mathbf{B}}_i + \frac{1}{2}\underline{\mathbf{B}}_{i+1}; i = 0, \ldots, n - 1$. The de Casteljau algorithm now becomes:

$$\underline{\mathbf{B}}_i^r(t) = (1 - t)\underline{\mathbf{B}}_i^{r-1}(t) + t\underline{\mathbf{B}}_{i+1}^{r-1}(t),$$

thus assuring that

$$\mathrm{cr}(\underline{\mathbf{B}}_i^{r-1}, \underline{\mathbf{B}}_i^r, \underline{\mathbf{Q}}_i^{r-1}, \underline{\mathbf{B}}_{i+1}^{r-1}) = \frac{t}{1 - t},$$

where we have set, in analogy to (7.8):

$$\underline{\mathbf{Q}}_i^r = (\underline{\mathbf{B}}_i^r \wedge \underline{\mathbf{B}}_{i+1}^r) \wedge (\underline{\mathbf{Q}}_i^{r-1} \wedge \underline{\mathbf{Q}}_{i+1}^{r-1}).$$

This is, of course, exactly the dual of the de Casteljau algorithm (7.7).

For control point curves, the tangent corresponding to t is given by $\underline{\mathbf{T}}(t) = \underline{\mathbf{b}}(t) \wedge \underline{\mathbf{b}}(t)$. For control line curves, we then have that the point

---

[9]This wording is misleading: both concepts are dual to each other — neither is favored over the other.

corresponding to t is given as the intersection of two lines, namely $\underline{\mathbf{p}}(t) = \underline{\mathbf{B}}(t) \wedge \underline{\dot{\mathbf{B}}}(t)$. In particular, the endpoints of the curve are given by $\underline{\mathbf{B}}_0 \wedge \underline{\mathbf{B}}_1$ and $\underline{\mathbf{B}}_{n-1} \wedge \underline{\mathbf{B}}_n$, respectively. We may also express the point on the curve in terms of the de Casteljau algorithm: $\underline{\mathbf{p}}(t) = \underline{\mathbf{B}}_0^{n-1}(t) \wedge \underline{\mathbf{B}}_1^{n-1}(t)$.

If we are given a control point curve, defined by a set of control points $\underline{\mathbf{b}}_0, \ldots, \underline{\mathbf{b}}_n$, how can we express it as a control line curve, with control lines $\underline{\mathbf{B}}_0, \ldots, \underline{\mathbf{B}}_m$?[10] Instead of treating the general case, we first restrict ourselves to the cubic case $n = 3$. Recall that the tangent $\underline{\mathbf{T}}(t)$ at $\underline{\mathbf{b}}_0^3(t)$ is spanned by the intermediate control points $\underline{\mathbf{b}}_0^2(t)$ and $\underline{\mathbf{b}}_1^2(t)$:

$$
\begin{aligned}
\underline{\mathbf{T}} &= \underline{\mathbf{b}}_0^2(t) \wedge \underline{\mathbf{b}}_1^2(t) \\
&= [(1-t)^2 \underline{\mathbf{b}}_0 + 2t(1-t)\underline{\mathbf{b}}_1 + t^2 \underline{\mathbf{b}}_2] \\
&\quad \wedge [(1-t)^2 \underline{\mathbf{b}}_1 + 2t(1-t)\underline{\mathbf{b}}_2 + t^2 \underline{\mathbf{b}}_3] \\
&= (1-t)^4 \underline{\mathbf{b}}_0 \wedge \underline{\mathbf{b}}_1 \\
&\quad + 2t(1-t)^3 \underline{\mathbf{b}}_0 \wedge \underline{\mathbf{b}}_2 \\
&\quad + t^2(1-t)^2 (\underline{\mathbf{b}}_0 \wedge \underline{\mathbf{b}}_3 + 3\underline{\mathbf{b}}_1 \wedge \underline{\mathbf{b}}_2) \\
&\quad + 2t^3(1-t)\underline{\mathbf{b}}_1 \wedge \underline{\mathbf{b}}_3 \\
&\quad + t^4 \underline{\mathbf{b}}_2 \wedge \underline{\mathbf{b}}_3.
\end{aligned}
$$

Thus we now have a quartic control line curve whose control lines are given by

$$
\begin{aligned}
&\underline{\mathbf{b}}_0 \wedge \underline{\mathbf{b}}_1 \\
&\tfrac{1}{2}\underline{\mathbf{b}}_0 \wedge \underline{\mathbf{b}}_2 \\
&\tfrac{1}{6}(\underline{\mathbf{b}}_0 \wedge \underline{\mathbf{b}}_3 + 3\underline{\mathbf{b}}_1 \wedge \underline{\mathbf{b}}_2) \\
&\tfrac{1}{2}\underline{\mathbf{b}}_1 \wedge \underline{\mathbf{b}}_3 \\
&\underline{\mathbf{b}}_2 \wedge \underline{\mathbf{b}}_3.
\end{aligned}
$$

Dually, we may convert a control line cubic to a control point quartic, using the same formalism as above.

In general, this conversion process will raise the degree from n to $2n-2$, considering that it involves terms of the form $\underline{\mathbf{T}}(t) = \underline{\mathbf{b}}_0^{n-1}(t) \wedge \underline{\mathbf{b}}_1^{n-1}(t)$ or $\underline{\mathbf{b}}(t) = \underline{\mathbf{B}}_0^{n-1}(t) \wedge \underline{\mathbf{B}}_1^{n-1}(t)$.

Let us now turn to the 3D case. The "usual" Bézier curves are again called control point curves; their duals are *control plane* curves of the form

$$
\underline{\mathbf{B}}(t) = \sum_{i=0}^{n} \underline{\mathbf{B}}_i(t), \tag{7.42}
$$

---

[10] Note that we do not know the degree m of the control line form yet!

but now with $\underline{\mathbf{B}}(t)$ as well as the $\underline{\mathbf{B}}_i$ being planes. The curve $\underline{\mathbf{B}}(t)$ is defined as the envelope of its *osculating planes*[11] $\underline{\mathbf{B}}(t)$. This statement follows from the following comparison of properties of control point and control plane curves:

control point curve:		control plane curve:	
point:	$\underline{\mathbf{b}}(t)$	osculating plane:	$\underline{\mathbf{B}}(t)$
tangent:	$\underline{\mathbf{b}}(t) \wedge \underline{\dot{\mathbf{b}}}(t)$	tangent:	$\underline{\mathbf{B}}(t) \wedge \underline{\dot{\mathbf{B}}}(t)$
osculating plane:	$< \underline{\mathbf{b}}, \underline{\dot{\mathbf{b}}}, \underline{\ddot{\mathbf{b}}} >$	point:	$< \underline{\mathbf{B}}, \underline{\dot{\mathbf{B}}}, \underline{\ddot{\mathbf{B}}} >$

Here, we have used the notation that $< \underline{\mathbf{a}}, \underline{\mathbf{b}}, \underline{\mathbf{c}} >$ defines a plane and that $< \underline{\mathbf{A}}, \underline{\mathbf{B}}, \underline{\mathbf{C}} >$ defines a point — see Section 1.14 for details.

We may again ask: given a control point curve, how do we express it as a control plane curve? Taking the cubic case as before, we have

$$
\begin{aligned}
\underline{\mathbf{B}}(t) &= \ < \underline{\mathbf{b}}_0^1(t), \underline{\mathbf{b}}_1^1(t), \underline{\mathbf{b}}_2^1(t) > \\
&= \ (1-t)^3 < \underline{\mathbf{b}}_0, \underline{\mathbf{b}}_1, \underline{\mathbf{b}}_2 > \\
&\quad +t(1-t)^2 < \underline{\mathbf{b}}_0, \underline{\mathbf{b}}_1, \underline{\mathbf{b}}_3 > \\
&\quad +t^2(1-t) < \underline{\mathbf{b}}_0, \underline{\mathbf{b}}_2, \underline{\mathbf{b}}_3 > \\
&\quad +t^3 < \underline{\mathbf{b}}_1, \underline{\mathbf{b}}_2, \underline{\mathbf{b}}_3 > \ .
\end{aligned}
$$

In the plane case, the dual of a control point cubic is a control line quartic — yet for 3D curves, the dual of a control point cubic is a control plane cubic! For higher degrees, however, we will again see the degrees going up: the dual of a degree n control point curve is a degree $3n - 3$ control plane curve.

The importance of these dual developments is in the theory of developable surfaces, which we shall discuss in Section 11.10. The key observation will be that while the planes $\underline{\mathbf{B}}(t)$ are osculating planes of a curve, they themselves, being a one parameter family of planes, envelope a surface. Such surfaces are called developable.

7.13 Hybrid Bézier Curves

It is frequently important to approximate a rational Bézier curve by an integral, i.e., polynomial one. This need will arise when a rational Bézier

---

[11]The osculating plane of a space curve is best explained as follows: fix a point on the curve, and also two more points "nearby." These three points define a plane. Let the two points approach the first point; in the limit, all three will coincide and the planes will converge to a limit plane — the osculating plane.

curve is produced in one CAD system, and is to be imported into another system, which can only handle integral curves. A direct conversion rational → polynomial is impossible, and so one must resort to approximation methods.

This section presents an approximation method due to Sederberg and Kakimoto [139], with an extension by Worsey, Sederberg, and Wang [161].

The basic idea is this: any rational Bézier curve

$$\mathbf{b}(t) = \frac{\sum_{i=0}^{n} w_i \mathbf{b}_i B_i^n(t)}{\sum_{i=0}^{n} w_i B_i^n(t)} \tag{7.43}$$

can be split into a polynomial and a rational part as follows. The polynomial part can be of arbitrary degree m. But one of the polynomial control points, $\mathbf{p}_s$, is given special treatment: the point $\mathbf{p}_s$ will not be constant; instead it will vary along a rational Bézier curve of the original degree n and with the original weights w_i. The hybrid form of a rational Bézier curve then takes the form

$$\mathbf{b}(t) = \sum_{\substack{i=0 \\ i \neq s}}^{m} \mathbf{p}_i B_i^m(t) + \frac{\sum_{i=0}^{n} w_i \mathbf{c}_i B_i^n(t)}{\sum_{i=0}^{n} w_i B_i^n(t)} \cdot B_s^m(t). \tag{7.44}$$

The numbers s and m are arbitrary — after they are selected, one has to determine the unknowns $\mathbf{p}_i$ and $\mathbf{c}_i$. We do this by equating the right hand sides of (7.43) and (7.44):

$$\sum_{i=0}^{n} w_i \mathbf{b}_i B_i^n(t) = \left(\sum_{i=0}^{n} w_i B_i^n(t)\right)\left(\sum_{\substack{i=0 \\ i \neq s}}^{m} \mathbf{p}_i B_i^m(t)\right) + \left(\sum_{i=0}^{n} w_i \mathbf{c}_i B_i^n(t)\right) B_s^m(t).$$
$$\tag{7.45}$$

The degree of the left hand side curve can be elevated from n to $n+m$. We make use of the *generalized degree elevation formula*:

$$\sum_{i=0}^{n} \mathbf{b}_i B_i^n(t) = \sum_{i=0}^{n+m} \mathbf{b}_i^{(m)} B_i^{n+m}(t)$$

with

$$\mathbf{b}_i^{(m)} = \sum_{j=0}^{n} \mathbf{b}_j \binom{n}{j} \frac{\binom{m}{i-j}}{\binom{n+m}{i}} = \sum_{j+k=i} \mathbf{b}_i \binom{n}{j} \frac{\binom{m}{k}}{\binom{n+m}{i}}.$$

We will also use the product formula

$$B_i^n(t) B_j^m(t) = \frac{\binom{n}{i}\binom{m}{j}}{\binom{n+m}{i+j}} B_{i+j}^{n+m}(t),$$

thus transforming (7.45) into

$$\sum_{i=0}^{n+m} \left(\sum_{j+k=i} w_j \mathbf{b}_j \frac{\binom{n}{j}\binom{m}{k}}{\binom{n+m}{i}} \right) B_i^{n+m}(t)$$

$$= \sum_{i=0}^{n+m} \left(\sum_{j+k=i} w_j \mathbf{p}_k \frac{\binom{n}{j}\binom{m}{k}}{\binom{n+m}{i}} \right) B_i^{n+m} + \sum_{i=s}^{s+n} w_{i-s} \mathbf{c}_{i-s} \frac{\binom{m}{s}\binom{n}{i-s}}{\binom{n+m}{i}} B_i^{n+m}(t).$$

For $i < s$, this simplifies enough so that we can compare coefficients:

$$\sum_{j+k=i} w_j \mathbf{b}_j \binom{n}{j}\binom{m}{k} = \sum_{j+k=i} w_j \mathbf{p}_k \binom{n}{j}\binom{m}{k}.$$

This allows us to compute the unknown $\mathbf{p}_i$ for $i = 0, \ldots, s-1$:

$$\mathbf{p}_0 = \mathbf{b}_0, \quad \mathbf{p}_1 = \mathbf{b}_0 + \frac{w_1 n}{w_0 m} \mathbf{b}_1 - \frac{w_1 n}{w_0 m} \mathbf{p}_0, \quad \text{etc.}$$

Analogously, we obtain $\mathbf{p}_m, \ldots, \mathbf{p}_{s+1}$ as

$$\mathbf{p}_m = \mathbf{b}_n, \quad \mathbf{p}_{m-1} = \mathbf{b}_n + \frac{w_{n-1} m}{w_n n} \mathbf{b}_{n-1} - \frac{w_{n-1} m}{w_n n} \mathbf{p}_m, \quad \text{etc.}$$

Having found all $\mathbf{p}_i$, we still need to determine the $\mathbf{c}_i$. For $i = s, \ldots, s+n$, we obtain

$$\sum_{j+k=i} w_j \mathbf{b}_j \binom{n}{j}\binom{m}{k} = \sum_{j+k=i} w_j \mathbf{p}_k \binom{n}{j}\binom{m}{k} + w_{i-s}\mathbf{c}_{i-s} \binom{m}{s}\binom{n}{i-s}$$

and thus

$$\mathbf{c}_{i-s} = \frac{\sum_{j+k=i} w_j \binom{n}{j}\binom{m}{k}[\mathbf{b}_j - \mathbf{p}_k]}{w_{i-s}\binom{m}{s}\binom{n}{i-s}}.$$

What is the benefit of this much algebra? Consider Fig. 7.17. A rational quartic (control polygon not shown) is written as an integral quadratic, with the control point $\mathbf{b}_1$ replaced by another rational quartic. Notice that the size of this rational quartic is considerably less than that of the original curve. Loosely speaking, the size of the rational component tells us how "rational" the original curve was: the smaller the size of the rational component, the closer the original curve was to being polynomial.

As we increase the degree m of the integral curve, the size of the rational component shrinks — as m tends to infinity, it will shrink to a point. This gives the desired algorithm for the approximation of a rational curve by a polynomial one: increase m until the size of the rational component

Figure 7.17. Hybrid curves: a rational quartic (control polygon not shown) is written as an integral quadratic with a variable center control point.

(conveniently measured in terms of its minmax box) shrinks below a given tolerance, then replace the rational component by the center of its minmax-box.

A. Worsey, T. Sederberg, and G. Wang [161] have streamlined the process of increasing the degree of the integral component. Instead of recomputing everything all over for each new value of m, one can capitalize on what has already been worked out. We discuss the case of a rational cubic, setting $m = 2$ and $s = 1$. We then have

$$\mathbf{b}(t) = \frac{\sum_{i=0}^{n} w_i \mathbf{b}_i B_i^n(t)}{\sum_{i=0}^{n} w_i B_i^n(t)} = \sum_{\substack{i=0 \\ i \neq 1}}^{2} \mathbf{p}_i^{(1)} B_i^2(t) + \frac{\sum_{i=0}^{3} w_i \mathbf{c}_i^{(1)} B_i^3(t)}{\sum_{i=0}^{n} w_i B_i^3(t)} B_1^2(t),$$

from which we get

$$\mathbf{p}_0^{(1)} = \mathbf{b}_0,$$

$$\mathbf{p}_2^{(1)} = \mathbf{b}_3,$$

$$\mathbf{c}_0^{(1)} = \mathbf{b}_0 + \frac{3w_1}{2w_0}[\mathbf{b}_1 - \mathbf{b}_0],$$

$$\mathbf{c}_1^{(1)} = \mathbf{b}_1 + \frac{1}{6w_1}[w_0(\mathbf{b}_0 - \mathbf{b}_3) + 3w_2(\mathbf{b}_2 - \mathbf{b}_0)],$$

$$\mathbf{c}_2^{(1)} = \mathbf{b}_2 + \frac{1}{6w_2}[w_3(\mathbf{b}_3 - \mathbf{b}_0) + 3w_1(\mathbf{b}_1 - \mathbf{b}_3)],$$

$$\mathbf{c}_3^{(1)} = \mathbf{b}_3 + \frac{3w_2}{2w_3}[\mathbf{b}_2 - \mathbf{b}_3].$$

The iteration now yields an expression for the rational component as yet another hybrid curve:

$$\mathbf{b}(t) = \sum_{\substack{i=0 \\ i \neq 1}}^{2} \mathbf{p}_i^{(1)} B_i^2(t) + \left[\sum_{\substack{i=0 \\ i \neq 1}}^{2} \mathbf{p}_i^{(2)} B_i^2(t) \frac{\sum_{i=0}^{3} w_i \mathbf{c}_i^{(2)} B_i^3(t)}{\sum_{i=0}^{n} w_i B_i^3(t)} \right] B_1^2(t). \quad (7.46)$$

The recursion proceeds by continually replacing the rational component of each representation by another hybrid form. It stops when the size of a rational component's minmax-box shrinks below a preset tolerance.

A different approach, based upon approximation, can be found in [13], [91], [92], [114].

7.14   Straight Lines as Bézier curves

If all control points of a rational Bézier curve are spaced evenly on a straight line $\mathbf{b}_0, \mathbf{b}_n$, and if the weights are unity, then the curve is in fact polynomial, tracing out the straight line in a linear fashion.

What if the control points are not evenly distributed? Let the control points be given by

$$\mathbf{b}_i = (1 - \alpha_i)\mathbf{b}_0 + \alpha_i \mathbf{b}_n; \quad i = 0, ..., n \tag{7.47}$$

where $\alpha_0 = 0, \alpha_n = 1$, i.e., they are distributed arbitrarily on a straight line. Can we find weights w_i such that $\mathbf{x}(t)$ traces out the line $\mathbf{b}_0, \mathbf{b}_n$ linearly? The answer is "yes"; the following argument is based on [59].

We can determine the unknown weights w_i from

$$\frac{\sum_{i=0}^n w_i \mathbf{b}_i B_i^n(t)}{\sum_{i=0}^n w_i B_i^n(t)} = (1 - t)\mathbf{b}_0 + t\mathbf{b}_n.$$

Since we can assume $\mathbf{b}_0 = \begin{bmatrix} 0 \\ 0 \end{bmatrix}$ and $\mathbf{b}_n = \begin{bmatrix} 1 \\ 0 \end{bmatrix}$, and consider only the x-component, we obtain

$$\frac{\sum_{i=0}^n w_i \alpha_i B_i^n(t)}{\sum_{i=0}^n w_i B_i^n(t)} = t$$

and hence

$$\sum_{i=0}^n w_i \alpha_i B_i^n(t) = t \sum_{i=0}^n w_i B_i^n(t).$$

Degree elevation and reindexing yields

$$\sum_{i=0}^{n+1} \left[\tfrac{i}{n+1} w_{i-1} \alpha_{i-1} + \tfrac{n+1-i}{n+1} w_i \alpha_i \right] B_i^{n+1}(t) = \sum_{i=0}^{n+1} \frac{i}{n+1} w_{i-1} B_i^{n+1}(t).$$

We now compare coefficients of B_i^{n+1} on both sides and obtain the recursion formula

$$w_i = \frac{i}{n+1-i} \frac{(1 - \alpha_{i-1})}{\alpha_i} w_{i-1}; \quad i = 1, ..., n; \ w_0 \neq 0. \tag{7.48}$$

A similar result holds if we do not consider control *points*, but a mix of control points and control *vectors*. This is outlined by Fiorot and Jeannin [69].

In general, one will not be faced by a situation as described by (7.47). But if a control polygon is given and it is not too far from linear, then (7.48) stands a good chance of producing "good" weights.

7.15   Problems

1. Sketch (manually) the curve from Example 7.1, including the point $\mathbf{b}(2)$.

2. Experiment graphically with control line curves. Compute your curves in projective space, then project into affine space for display.

3. If we take successive derivatives of a polynomial curve, eventually they are all zero. What are the degrees of successive derivatives of rational Bézier curves?

4. Equation (7.23) describes a hyperbola. It may be written in parametric form with control points $(0,0), (\alpha, \beta), (1,1)$. If in standard form, discuss how the middle weight and α, β depend on c.

5. Derive expressions for higher derivatives of a rational Bézier curve at $t = 0$, thus generalizing (7.29).

6. Experiment numerically with the two given versions of the de Casteljau algorithm: compare the completely rational version with the one that carries out all computations in projective space. In order to detect different behaviors, you will need to use rapidly varying weights.

7. The osculants of $\underline{\mathbf{o}}_l$ and $\underline{\mathbf{o}}_r$ of a rational cubic are conics. Is it possible that both are ellipses or both are hyperbolas?

8

Rational Cubics

Rational cubics are the simplest and most fundamental of all rational space curves. They are therefore important enough to warrant a whole chapter dedicated to their study alone. An excellent treatment of rational cubics is the one by Boehm[22]; another good reference is Forrest [77]. The more theoretically inclined reader will find valuable material in older books, such as [9] or [115].

8.1 Rational Cubic Conics

While rational cubics are space curves in general, an important subclass are those rational cubics that are actually conics, and hence planar. In projective space, we may obtain a cubic from a quadratic (a conic) by the process of degree elevation.

In $I\!\!P^2$, consider Fig. 8.1.[1] Let $\underline{b}_0, \underline{b}_1, \underline{b}_2$ be the original control points of the conic, and let $\underline{c}_0, \underline{c}_1, \underline{c}_2, \underline{c}_3$ be the control points of the corresponding cubic. They are related by

---

[1]When dealing with conics, there is obviously no need to go to $I\!\!P^3$!

$$
\begin{aligned}
\underline{c}_0 &= \underline{b}_0, \\
\underline{c}_1 &= \frac{1}{3}\underline{b}_0 + \frac{2}{3}\underline{b}_1, \\
\underline{c}_2 &= \frac{2}{3}\underline{b}_1 + \frac{1}{3}\underline{b}_2, \\
\underline{c}_3 &= \underline{b}_2.
\end{aligned}
$$

In affine space, the corresponding relationships are given by

$$
\begin{aligned}
c_0 &= \frac{b_0}{z_0}, \\
c_1 &= \frac{b_0 + 2b_1}{z_0 + 2z_1}, \\
c_2 &= \frac{2b_1 + b_2}{2z_1 + z_2} \\
c_3 &= \frac{b_2}{z_2},
\end{aligned}
$$

spelling out the more general equation (7.16). The weights w_i of the cubic are then given by

$$
\begin{aligned}
w_0 &= z_0, \\
w_1 &= z_0 + 2z_1, \\
w_2 &= 2z_1 + z_2 \\
w_3 &= z_2.
\end{aligned}
$$

There is an interesting interplay between the quadratic and cubic forms. In projective space, we have

$$
\underline{s} = \underline{b}(\tfrac{1}{2}) = \underline{c}(\tfrac{1}{2}) = [\underline{c}_0 \wedge \underline{c}_2] \wedge [\underline{c}_3 \wedge \underline{c}_1]. \tag{8.1}
$$

For a proof, we write

$$
\underline{s} = (1 - \alpha)\underline{b}_0 + \alpha(\tfrac{2}{3}\underline{b}_1 + \tfrac{1}{3}\underline{b}_2) = (1 - \beta)\underline{b}_2 + \beta(\tfrac{1}{3}\underline{b}_0 + \tfrac{2}{3}\underline{b}_1),
$$

from which deduce $\alpha = \beta = \tfrac{3}{4}$. Thus

$$
\underline{s} = \frac{1}{4}\underline{b}_0 + \frac{1}{2}\underline{b}_1 + \frac{1}{4}\underline{b}_2 = \underline{c}(\tfrac{1}{2}).
$$

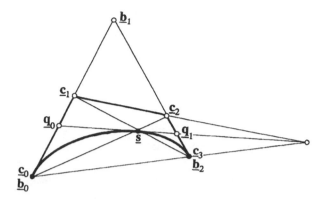

Figure 8.1. Cubic conics: the geometry of the quadratic $\to$ cubic degree elevation process in $\mathbb{P}^2$.

Let $\underline{\mathbf{q}}_0$ and $\underline{\mathbf{q}}_1$ be the intersections of the shoulder tangent $\underline{\mathbf{T}}(\frac{1}{2})$ with the quadratic control polygon, as shown in Fig. 8.1. Since $\underline{\mathbf{q}}_0 = \frac{1}{2}\underline{\mathbf{b}}_0 + \frac{1}{2}\underline{\mathbf{b}}_1$ and $\underline{\mathbf{q}}_1 = \frac{1}{2}\underline{\mathbf{b}}_1 + \frac{1}{2}\underline{\mathbf{b}}_2$, we find that

$$\mathrm{cr}(\underline{\mathbf{b}}_0, \underline{\mathbf{q}}_0, \underline{\mathbf{c}}_1, \underline{\mathbf{b}}_1) = \mathrm{cr}(\underline{\mathbf{b}}_1, \underline{\mathbf{q}}_1, \underline{\mathbf{c}}_2, \underline{\mathbf{b}}_2) = \frac{1}{2},$$

i.e., each set of four points is *harmonic*. This relationship holds true, of course, for the corresponding affine points. We also note that the lines $\underline{\mathbf{c}}_1 \wedge \underline{\mathbf{c}}_2$, $\underline{\mathbf{q}}_0 \wedge \underline{\mathbf{q}}_1$, and $\underline{\mathbf{b}}_0 \wedge \underline{\mathbf{b}}_2$ are concurrent, all passing through the point $\underline{\mathbf{b}}_2 - \underline{\mathbf{b}}_0$.

Let us define $\underline{\mathbf{p}}_i = \frac{1}{2}\underline{\mathbf{c}}_i + \frac{1}{2}\underline{\mathbf{c}}_{i+1}$; $i = 0, 1, 2$. (These points will be mapped to the affine weight points of the rational cubic.) It is not hard to show that the three points $\underline{\mathbf{b}}_1, \underline{\mathbf{p}}_1$, and $\underline{\mathbf{s}}$ are collinear and that the line $\underline{\mathbf{p}}_0 \wedge \underline{\mathbf{p}}_2$ passes through the point $\underline{\mathbf{b}}_2 - \underline{\mathbf{b}}_0$.

We now consider the following affine problem: Given four points c_0, c_1, c_2, c_3 forming a convex and planar polygon, is it possible to find weights w_i such that the resulting rational cubic is actually a conic? The answer is "yes", and the weights can be found as follows:

First, we find the intersection b_1 of the first and last polygon legs. Then q_0, a conic weight point, is determined by the requirement that $\mathrm{cr}(c_0, q_0, c_1, b_1) = \frac{1}{2}$. Setting $z_0 = 1$, we then have

$$z_1 = \mathrm{ratio}(c_0, q_0, b_1) = \frac{1}{2}\mathrm{ratio}(c_0, c_1, b_1).$$

Here, z_1 is a conic weight. Proceeding in the same way again, we find that

$$z_2 = z_1 \mathrm{ratio}(\mathbf{b}_1, \mathbf{q}_1, \mathbf{c}_3) = \frac{1}{2} z_1 \mathrm{ratio}(\underline{\mathbf{b}}_1, \mathbf{c}_2, \mathbf{c}_3).$$

The three weights $1, z_1, z_2$ are the weights of the conic; degree elevation now gives the desired cubic weights.

This method shows how we could estimate "good" weights for a rational cubic if only its control polygon were given and the weights had to be made up. The method does not work, understandably, for nonconvex or nonplanar polygons.

The technique may also be viewed as one for "degree reduction" for rational cubics with convex control polygons: if a rational cubic is given and a rational quadratic approximation is desired, its weights are z_0, z_1, z_2 and the control polygon is $\mathbf{c}_0, \mathbf{b}_1, \mathbf{c}_3$.

8.2   Embedding Cones

For every rational cubic, we can find a quadratic cone such that it "winds around" that cone. A quadratic cone is defined as follows: it is a surface with the three properties that

  a) the tangent plane to any point of the cone has a whole line in common with the cone,

  b) all those lines are concurrent,

  c) any planar section of the cone is a conic, i.e, a quadratic curve.

Before we elaborate further, it should be noted that any space curve $\underline{\mathbf{x}}(t)$ lies on a cone: just pick any point $\mathbf{p}$ and "cone off", i.e., form lines $\underline{\mathbf{p}} \wedge \underline{\mathbf{x}}(t)$. These lines envelope a conical surface. We will show here that for the special case of parametric cubics, we can find cones which are quadratic.

We will investigate this phenomenon in projective space $I\!\!P^3$. For a given cubic, define two *Ball points*:

$$\underline{\mathbf{l}}_1 = \frac{3}{2}\mathbf{b}_1 - \frac{1}{2}\mathbf{b}_0, \quad \underline{\mathbf{l}}_2 = \frac{3}{2}\mathbf{b}_2 - \frac{1}{2}\mathbf{b}_3, \tag{8.2}$$

after A. Ball, who introduced these points for use in his design system CONSURF, see [10], [11], [12].

We may write our cubic using the Ball points:

$$\underline{\mathbf{b}}(t) = (1-t)^2 \underline{\mathbf{b}}_0 + 2(1-t)^2 t \underline{\mathbf{l}}_1 + 2t^2(1-t)\underline{\mathbf{l}}_2 + t^2 \underline{\mathbf{b}}_3. \tag{8.3}$$

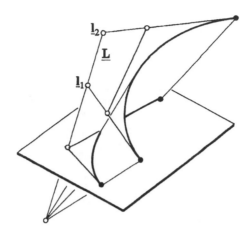

Figure 8.2. Projecting a cubic: a cubic is projected into a plane such that the two Ball points have identical images.

This is proved by simply inserting $\underline{l}_i$'s definition into the cubic Bézier form. In the special case that both $\underline{l}_i$ agree: $\underline{l} := \underline{l}_1 = \underline{l}_2$, our cubic is actually a quadratic, with control polygon $\underline{b}_0, \underline{l}, \underline{b}_2$.

Assuming from now on that $\underline{l}_1 \neq \underline{l}_2$, let $\underline{P}$ be any plane not containing the *Ball line* $\underline{L} = \underline{l}_1 \wedge \underline{l}_2$ and let $\underline{v}$ be a point on that line. Let us now project our cubic into $\underline{P}$ through $\underline{v}$. Then the images of the two Ball points coincide, yielding a planar image control polygon $\hat{\underline{b}}_0, \hat{\underline{l}}, \hat{\underline{b}}_3$, see Fig. 8.2. It defines a conic $\underline{c}(t)$ in the plane $\underline{P}$ which is the projection of the original cubic. As every line $\underline{c}(t) \wedge \underline{b}(t)$ passes through $\underline{v}$, we find that the curve $\underline{b}(t)$ is on a quadratic cone with generating lines $\underline{c}(t) \wedge \underline{b}(t)$ and with vertex $\underline{v}$. Note that the Ball line $\underline{L}$ itself does not lie on the cone; it intersects it at the point $\underline{v}$, the cone's vertex.

Different choices of $\underline{v}$ on $\underline{L}$ will give rise to different cones, and so there is in fact a one-parameter family of cones that $\underline{b}(t)$ lies on. Of all points on $\underline{L}$, one is special with respect to the given cubic: it is the difference of the Ball points:

$$
\begin{aligned}
\underline{l}_1 - \underline{l}_2 &= \frac{1}{2}(3\underline{b}_1 - \underline{b}_0 - 3\underline{b}_2 + \underline{b}_3) \\
&= \frac{1}{2}\Delta^3 \underline{b}_0 \\
&\doteq \underline{b}(\infty) \\
&\doteq \dot{\underline{b}}(\infty) \\
&\doteq \ddot{\underline{b}}(\infty).
\end{aligned}
$$

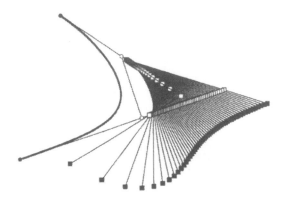

Figure 8.3. Ball lines: depending on its parametrization, a cubic possesses different Ball lines. Ball points are marked by hollow squares, their differences are marked by solid squares.

Thus the point $\underline{\mathbf{b}}(\infty)$ lies on $\underline{\mathbf{L}}$.

If we reparametrize the cubic using a Moebius transformation, the Ball points will change; but again $l_1 - l_2$ will be on the cubic! Fig. 8.3 shows how the Ball line is affected by reparametrizations.

As we project our cubic into affine space, we obtain a rational cubic with Ball points

$$l_1 = \frac{3w_1\mathbf{b}_1 - w_0\mathbf{b}_0}{3w_1 - w_0}, l_2 = \frac{3w_2\mathbf{b}_2 - w_3\mathbf{b}_3}{3w_2 - w_3}. \tag{8.4}$$

As cones are mapped to cones by projections, the rational cubic also lies on a one-parameter family of cones. The Ball line $\mathbf{L}$ again intersects the rational cubic for $t = \infty$. In the case of an integral cubic, the cone becomes a *parabolic cylinder*, and $\mathbf{L}$ is called an *asymptotic direction*, meaning that for $t \to \infty$, the curve will have tangents parallel to $\mathbf{L}$. This is easily seen by observing that the dominating term in $\lim_{t \to \infty} \dot{\mathbf{b}}(t)$ is given by $\Delta^3\mathbf{b}_0$ which equals $l_2 - l_1$. Fig. 8.4 illustrates the integral case.

8.3   The Embedding Bilinear Patch

Let a projective cubic Bézier curve be given by its control points $\underline{\mathbf{b}}_0, \underline{\mathbf{b}}_1, \underline{\mathbf{b}}_2, \underline{\mathbf{b}}_3$ we define

$$\underline{\mathbf{a}}_1 = \frac{3}{2}\underline{\mathbf{b}}_1 - \frac{1}{2}\underline{\mathbf{b}}_3, \quad \underline{\mathbf{a}}_2 = \frac{3}{2}\underline{\mathbf{b}}_2 - \frac{1}{2}\underline{\mathbf{b}}_0. \tag{8.5}$$

Figure 8.4. Integral cubics: the curve lies on a parabolic cylinder. The cylinder axis is parallel to the Ball line.

Then the four points $\underline{\mathbf{b}}_0, \underline{\mathbf{a}}_1, \underline{\mathbf{a}}_2, \underline{\mathbf{b}}_3$ form a bilinear surface (see Section 11.1 for details):

$$\underline{\mathbf{x}}(u, v) = \underline{\mathbf{b}}_0(1 - u)(1 - v) + \underline{\mathbf{b}}_3 u(1 - v) + \underline{\mathbf{a}}_1(1 - u)v + \underline{\mathbf{a}}_2 uv.$$

This surface is a doubly ruled quadric, a so-called annular quadric. To show that our cubic lies on this surface, we set

$$v = B_1^2(u) = 2(1 - u)u$$

and insert this into (8.3): we recover the equation of the cubic! See Fig. 8.5 for an illustration.

After we project (8.3) into affine space, the projective bilinear surface becomes an affine rational bilinear patch. The four corners of this patch are given by the projections $\mathbf{b}_i$ and $\mathbf{a}_i$ of the $\underline{\mathbf{b}}_i$ and the $\underline{\mathbf{a}}_i$. We note that

$$\mathbf{a}_1 = \frac{3w_1\mathbf{b}_1 - w_3\mathbf{b}_3}{3w_1 - w_3} \quad \text{and} \quad \mathbf{a}_2 = \frac{3w_2\mathbf{b}_2 - w_0\mathbf{b}_0}{3w_2 - w_0}.$$

Together with the result from Section 8.2, we see that every rational cubic is the intersection between a cone and a bilinear patch. For the integral case, when all four w_i are equal, the embedding bilinear patch is a *hyperbolic paraboloid*.

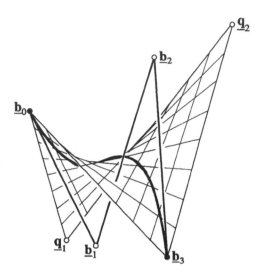

Figure 8.5. The embedding bilinear patch: every cubic lies entirely on a doubly ruled quadric.

8.4   Asymptotes

A cubic[2] $\underline{\mathbf{b}}(t)$ in $I\!\!P^3$ intersects every plane at least once. In particular, it intersects the plane $\underline{\mathbf{H}}$ which will be mapped to the plane at infinity when we project into affine space. Let $\underline{\mathbf{x}}$ be such a point of intersection; also let $\underline{\mathbf{T}}$ be the cubic's tangent at $\underline{\mathbf{x}}$. As we project into affine space, $\underline{\mathbf{x}}$ will be mapped to a point at infinity, and $\underline{\mathbf{T}}$ will be mapped to an *asymptote*. An asymptote is thus the curve's tangent at infinity. The reader is invited to compare this situation with the similar case of hyperbolas, see Fig. 3.16.

What happens when $\underline{\mathbf{H}}$ is *tangent* to the cubic? Then the resulting (affine) asymptote is itself at infinity, and the rational cubic touches the plane at infinity. Note that there are two different cases: the projective cubic may touch $\underline{\mathbf{H}}$ (just as $y = (1-x)^3$ touches the $x-$axis at $x = 1$) or it may intersect $\underline{\mathbf{H}}$ (just as $y = x^3$ intersects the $x-$axis at $x = 0$).

Algebraically speaking, the presence of an asymptote means that the denominator $w(t)$ of the rational cubic $\mathbf{b}(t)$ has at least one real zero t_0; as $t \to t_0$, the curve $\mathbf{b}(t)$ tends to infinity. Fig. 8.6 illustrates the geometry. Be sure you understand how the parameter t traces out the curve!

We could now reparametrize our rational cubic such that we map $0, 1, t_0$ to $0, 1, \infty$. Then the curve would tend to infinity as the new parameter s

---

[2]Here, we refer to actual cubics, not quadratics who were made cubic by degree elevation.

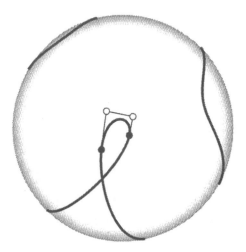

Figure 8.6. Asymptotes: the denominator has three real zeroes — the curve has three asymptotes.

tends to infinity, thus resembling the behavior of a parabola in standard form (Sec. 5.2). Since

$$\lim_{s \to \infty} \mathbf{b}(s) = \frac{\Delta^3[\hat{w}_0 \mathbf{b}_0]}{\Delta^3 \hat{w}_0},$$

it follows that $\Delta^3 \hat{w}_0 = 0$, meaning that the new denominator is actually quadratic. Recalling that $\hat{w}_i = c^i w_i$ for some constant c, we can find c from

$$\Delta^3 \hat{w}_0 = w_0 - 3c w_1 + 3c^2 w_2 - c^3 w_3 = 0.$$

Note that this cubic equation for c always has (at least) one real solution, provided that the denominator was actually cubic.

With this new set of weights, the denominator can be written as a quadratic, and our rational cubic becomes

$$\mathbf{b}(s) = \frac{\sum_{i=0}^{3} \hat{w}_i \mathbf{b}_i B_i^3(s)}{\sum_{i=0}^{2} v_i B_i^2(s)}, \tag{8.6}$$

where the v_i are obtained from degree reducing the denominator. This form of a rational cubic was first considered by L. Cremona in 1859. It was named *T-conic* by M. Rowin [131] in 1964, quoted from R. Forrest [75].

Writing a rational cubic in T-conic form simplifies the form of the asymptotes. We know that one asymptote (or asymptotic direction) is given by the Ball line $\mathbf{L}$.[3] There may be (at most) two more, and they are then given by the zeroes of the numerator $\sum_{i=0}^{2} v_i B_i^2(s)$. It would be useful to detect if these zeroes are in the interval $[0, 1]$. No such zero exists if all three v_i are positive. If v_0 and v_2 are positive, then the condition $v_1 > -\sqrt{v_0 v_2}$ ensures that no zero is in $[0, 1]$ (see Boehm [22]).

8.5 The Hermite Form

In affine geometry, *Hermite interpolation* is a well-known way to interpolate to curves when derivative information is given. See [38] or [57] for details. Here, we shall derive this curve form from a projective formulation.

Starting with a projective cubic Bézier curve, let us introduce $\underline{\mathbf{t}}_0 = \underline{\mathbf{b}}_1 - \underline{\mathbf{b}}_0$ and $\underline{\mathbf{t}}_1 = \underline{\mathbf{b}}_3 - \underline{\mathbf{b}}_2$. We can then write the original cubic as

$$\underline{\mathbf{b}}(t) = \underline{\mathbf{b}}_0 B_0^3(t) + [\underline{\mathbf{t}}_0 + \underline{\mathbf{b}}_0] B_1^3(t) + [\underline{\mathbf{b}}_3 - \underline{\mathbf{t}}_1] B_2^3(t) + \underline{\mathbf{b}}_3 B_3^3(t).$$

This may be reformulated as

$$\underline{\mathbf{b}}(t) = \underline{\mathbf{b}}_0 H_0^3(t) + 3\underline{\mathbf{t}}_0 H_1^3(t) + 3\underline{\mathbf{t}}_1 H_2^3(t) + \underline{\mathbf{b}}_3 H_3^3(t)$$

by setting

$$H_0^3 = B_0^3 + B_1^3, H_1^3 = \frac{1}{3} B_1^3, H_2^3 = -\frac{1}{3} B_2^3, H_3^3 = B_2^3 + B_3^3.$$

Note that $\underline{\dot{\mathbf{b}}}(0) \hat{=} \underline{\mathbf{t}}_0$ and $\underline{\dot{\mathbf{b}}}(1) \hat{=} \underline{\mathbf{t}}_1$ — but don't forget that $\underline{\mathbf{b}}(0)$ and $\underline{\mathbf{b}}(1)$ are points in $I\!\!P^3$!

Assuming that $\underline{\mathbf{b}}(t)$ is in standard form, i.e., $w_0 = w_3 = 1$, we project into affine space to obtain the rational form

$$\mathbf{b}(t) = \frac{\mathbf{b}_0 H_0^3(t) + 3v_0 \mathbf{t}_0 H_1^3(t) + 3v_1 \mathbf{t}_1 H_2^3(t) + \mathbf{b}_3 H_3^3(t)}{H_0^3(t) + 3v_0 H_1^3(t) + 3v_1 H_2^3(t) + H_3^3(t)}, \tag{8.7}$$

with v_0 and v_1 as the fourth coordinates of $\underline{\mathbf{t}}_0$ and $\underline{\mathbf{t}}_1$, respectively.

In affine space, derivatives correspond to vectors. If we want the projections of $\underline{\mathbf{t}}_0$ and $\underline{\mathbf{t}}_1$ to be derivatives of the rational curve $\mathbf{b}(t)$, then we must arrange for the projection to satisfy a special condition: it must map the line $\underline{\mathbf{H}} = \underline{\mathbf{t}}_0 \wedge \underline{\mathbf{t}}_1$ to an affine line at infinity. Then $v_0 = v_1 = 0$, and the $\underline{\mathbf{t}}_i$ are mapped to *tangent vectors* $\mathbf{t}_i$.[4] Since $H_0^3(t) + H_3^3(t) \equiv 1$, the rational curve (8.7) now takes on the simple polynomial form

$$\mathbf{b}(t) = \mathbf{b}_0 H_0^3(t) + 3\mathbf{t}_0 H_1^3(t) + 3\mathbf{t}_1 H_2^3(t) + \mathbf{b}_3 H_3^3(t), \tag{8.8}$$

---

[3]The Ball line changes as we reprametrize to get to the T-conic!

[4]Compare with the control vector approach from Section 7.11!

Figure 8.7. The Hermite form: left, as a polynomial in projective space, right: as a polynomial in affine space.

which is the familiar cubic Hermite interpolant. We easily verify that $\dot{\mathbf{b}}(0) = \mathbf{t}_0$ and $\dot{\mathbf{b}}(1) = \mathbf{t}_1$. Fig. 8.7 gives an illustration.

We see that the Bézier form is easily taken from projective space, where it is just a polynomial curve, to affine space, where it is rational polynomial. This is because all of its coefficients are *points*. The Hermite form, with its affine point/vector mix of coefficients, does not allow this simple treatment.

8.6   Cubic Lagrange Interpolation

In Section 4.7, we interpolated to four coplanar points and corresponding parameter values. With cubics, we can move to projective three-space and interpolate to five points with corresponding parameter values. Thus we are given five points $\mathbf{x}_0, \ldots, \mathbf{x}_4$ and corresponding parameter values $t_0, \ldots, t_4$. We want a cubic $\underline{\mathbf{c}}(t)$ such that $\underline{\mathbf{c}}(t_i) \hat{=} \underline{\mathbf{x}}_i$.

We closely follow Section 4.7, and only give the main steps here. Accordingly, we write a point on the cubic as

$$\underline{\mathbf{c}}(t) = [B_0^3(t), B_1^3(t), B_2^3(t), B_3^3(t)] \begin{bmatrix} \underline{\mathbf{b}}_0 \\ \underline{\mathbf{b}}_1 \\ \underline{\mathbf{b}}_2 \\ \underline{\mathbf{b}}_3 \end{bmatrix} . \tag{8.9}$$

We will first consider the four interpolation conditions $w_i \underline{\mathbf{x}}_i = \underline{\mathbf{c}}(t_i); i = 0, 1, 2, 3$, with unknown constants w_i. This leads to a 4×4 linear system

$$\boxed{\mathbf{x}} = B \boxed{\mathbf{b}}$$

with coefficient matrix $B = \{B_i^3(t_j)\}_{i,j=0}^3$ and solution $\boxed{\mathbf{b}} = B^{-1}\boxed{\mathbf{x}}$. The vector $\boxed{\mathbf{x}}$ still contains the unknowns w_i. Bringing in the fifth point $\mathbf{x}_4$, we obtain $\mathbf{x}_4 = B_4\boxed{\mathbf{b}}$ with $B_4 = [B_0^3(t_4), B_1^3(t_4), B_2^3(t_4), B_3^3(t_4)]$. Thus we obtain

$$\mathbf{x}_4 = B_4 B^{-1}\boxed{\mathbf{x}},$$

which is a 4×4 linear system for the unknowns w_i. Once we solved for them, there is the possibility that some of them may be negative!

8.7 Osculatory Interpolation

With rational cubics, it is easy to solve an interesting kind of interpolation problem: given a Bézier polygon $\mathbf{b}_0, \mathbf{b}_1, \mathbf{b}_2, \mathbf{b}_3$ and a curvature value at each endpoint, find a set of weights w_0, w_1, w_2, w_3 such that the corresponding rational cubic assumes the given curvatures at $\mathbf{b}_0$ and $\mathbf{b}_1$. The following method is very similar to one developed by T. Goodman in 1988; see [82]. We assume without loss of generality that $w_0 = w_3 = 1$.[5] The given curvatures κ_0 and κ_3 are given by

$$\kappa_0 = \frac{4}{3}\frac{w_2}{w_1^2}c_0, \quad \kappa_3 = \frac{4}{3}\frac{w_1}{w_2^2}c_1, \tag{8.10}$$

where

$$c_0 = \frac{\mathrm{area}[\mathbf{b}_0, \mathbf{b}_1, \mathbf{b}_2]}{\mathrm{dist}^3[\mathbf{b}_0, \mathbf{b}_1]}, \quad c_1 = \frac{\mathrm{area}[\mathbf{b}_1, \mathbf{b}_2, \mathbf{b}_3]}{\mathrm{dist}^3[\mathbf{b}_2, \mathbf{b}_3]}.$$

Equations(8.10) decouple nicely, so that we can determine our unknowns w_1 and w_2:

$$w_1 = \frac{4}{3}\Big[\frac{c_0^2}{\kappa_0^2}\frac{c_1}{\kappa_1}\Big]^{\frac{1}{3}}, \quad w_2 = \frac{4}{3}\Big[\frac{c_0}{\kappa_0}\frac{c_1^2}{\kappa_1^2}\Big]^{\frac{1}{3}}. \tag{8.11}$$

For planar control polygons, the quantities c_0 or c_1 may be negative — this happens when a control polygon is S-shaped. This is meaningful since curvature may be defined as *signed curvature* for 2D curves. Of course, one should then also prescribe the corresponding κ_0 and κ_1 as being negative, so that one ends up with positive weights.

A similar interpolation problem was addressed by Klass [95] and de Boor, Hollig, and Sabin for the nonrational case: they prescribe two points and corresponding tangent directions and curvatures [41]. The solution (when it exists) can only be obtained using an iterative method.

---

[5]Goodman [82] assumes that $w_1 = w_2 = 1$.

8.8   Problems

1. In Section 8.3, we remarked that every rational cubic is the intersection between a cone and a bilinear patch. How about the converse: is every intersection between a cone and a bilinear patch a rational cubic? (This question leads into the area of algebraic geometry. Pointers to the literature: [1], [2], [3],[4].)

2. If a rational cubic is *closed*, does this mean that it is a conic?

3. What is the model of a cubic in the spherical or circular models of the projective plane? If you have access to the appropriate kind of graphics equipment, illustrate!

4. A rational cubic may have one or three asymptotes, but not two. Why?

9

NURBS

NURBS is short for "**N**on–**U**niform **R**ational **B**-splines. This is a somewhat unlucky naming convention, since B-splines are non-uniform to begin with. We give an outline of B-spline curves in $I\!P^3$ — there, nothing is rational but rather piecewise polynomial. The "rational" part, as usual, comes when we project into affine space.

A B-spline curve is a curve consisting of several polynomial pieces (the *segments*, much like the conic splines in $I\!P^2$ from Chapter 6). The main difference is that now we do not restrict the degree — the only stipulation is that all segments be of the same degree. We will develop an algorithm for these curves — the de Boor algorithm — that is the analogue to the de Casteljau algorithm for Bézier curves.

A B-spline curve is defined by the degree n of each of its polynomial pieces (the segments), and by the number L of segments. Also needed is the *domain* of the curve, which is a partition of part of the real line into L intervals. This partition is a nondecreasing sequence of *knots* u_i. As a parameter u traces out the domain, the corresponding point $\underline{d}(u)$ on the B-spline curve traces out a piecewise polynomial curve, defined by a set of *control points* $\underline{d}_i$. The smoothness of this curve depends on the precise makeup of the knot sequence.

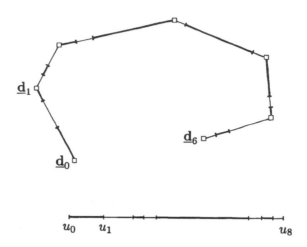

Figure 9.1. Engraving the knots: an example for $n = 3$.

9.1 Knot insertion

The most basic operation for a B-spline curve is *knot insertion*, a procedure
to add one knot to the knot sequence and change the control points such
that the (as yet undefined) curve will remain unchanged.

We are given a nondecreasing knot sequence $u_0, \ldots, u_K$ and the degree
n of a (to be defined) B-spline curve. The elements of the knot sequence,
the knots, may occur repeatedly, but no more than n times. Thus for $n = 2$,
the knot sequence $0, 1, 1, 2, 3$ is legal, but $0, 1, 1, 1, 2, 3$ is not. If a knot is
repeated k times in the knot sequence, it is said to have *multiplicity k*.

The curve will be defined by a control polygon

$$\mathbf{P} = \underline{\mathbf{d}}_0 \underline{\mathbf{d}}_1 \cdots \underline{\mathbf{d}}_p,$$

with $p = K - n + 1$. Thus there are as many control points as there are
successive n-tuples of knots. Example: Take a cubic B-spline curve and the
knot sequence $0, 1, 2, 3, 4, 5, 6, 7$. The 3-tuples (or triples, for the cubic case)
are $(0, 1, 2)$, $(1, 2, 3)$, etc., for a total of six triples. Thus a cubic B-spline
curve over this knot sequence will have six control points.

Take all spans of n successive intervals and map them to the control
polygon legs by projective maps:

$$[u_i, u_{i+n}] \to \underline{\mathbf{d}}_i \underline{\mathbf{d}}_{i+1}.$$

This way, each control polygon leg is "engraved" with a span of the knot
sequence.

Now let $\lambda = [u_I, u_{I+1}]$ be one of the intervals defined by two successive and different knots u_I and u_{I+1} for some I. Most of our constructions will be local, i.e., will only need intervals "near" λ. We will only consider intervals λ that are part of n successive spans — so not all intervals at the beginning and end of the knot sequence will qualify. It is easy to see that we must have $n - 1 \leq I \leq K - n$. A qualifying interval λ — called a *domain interval* — will thus be engraved on n successive polygon legs. We call the corresponding control polygon $\underline{\mathbf{P}}^\lambda$ and its control points $\underline{\mathbf{d}}_i^\lambda$:

$$\underline{\mathbf{P}}^\lambda = \underline{\mathbf{d}}_0^\lambda, \ldots, \underline{\mathbf{d}}_n^\lambda = \underline{\mathbf{d}}_{I-n+1}, \ldots, \underline{\mathbf{d}}_{I+1}.$$

See Fig. 9.1 for an illustration. We use the superscript λ when numbering relative to the interval λ.

There are n spans containing λ. We denote them by $\lambda_i^n; i = 1, \ldots, n$. Each of these spans is mapped to a control polygon leg by a projective map Φ_i^n:

$$\Phi_i^n : \lambda_i^n \rightarrow \underline{\mathbf{d}}_{i-1}^\lambda \underline{\mathbf{d}}_i^\lambda, \quad i = 1, \ldots, n.$$

Let $u \in \lambda$, where λ denotes a domain interval. Then each of the affine maps Φ_i^n takes u to a point on a control polygon leg, and we define

$$\underline{\mathbf{d}}_i^1(u) = \Phi_i^n(u); \quad i = 1, \ldots, n.$$

Since projective maps correspond to linear interpolation,[1] we may also write

$$\underline{\mathbf{d}}_i^1(u) = \frac{u - u_I}{u_{I+1} - u_I} \underline{\mathbf{d}}_{i-1} + \frac{u_r - u}{u_{I+1} - u_I} \underline{\mathbf{d}}_i. \tag{9.1}$$

We have augmented the knot sequence by one knot u, and we have augmented the control polygon $\underline{\mathbf{P}}^\lambda$ to a new polygon $\underline{\mathbf{P}}[u]$:

$$\underline{\mathbf{P}}[u] = \underline{\mathbf{d}}_0, \underline{\mathbf{d}}_0^1(u), \ldots \underline{\mathbf{d}}_n^1(u), \underline{\mathbf{d}}_n;$$

We call this process *knot insertion*, after W. Boehm [57]. See Fig. 9.2 for an illustration.

Using the augmented knot sequence and the appropriately adjusted control polygon, we may insert another knot v into the knot sequence and obtain an augmented polygon $\underline{\mathbf{P}}^\lambda[u, v]$. Conversely, we might have inserted v first and then u, resulting in $\underline{\mathbf{P}}^\lambda[v, u]$. We wish to show that the order of knot insertion does not matter, i.e.,

$$\underline{\mathbf{P}}^\lambda[u, v] = \underline{\mathbf{P}}^\lambda[v, u].$$

Consulting Fig. 9.3, we see that this amounts to showing that $\underline{\mathbf{d}}_1^1[u, v] = \underline{\mathbf{d}}_1^1[v, u]$. This, however, is a straightforward consequence of Menelaos' theorem from Section 1.8.

---

[1] Keep in mind that this statement applies to projective geometry; in affine geometry, projective maps correspond to rational linear interpolation.

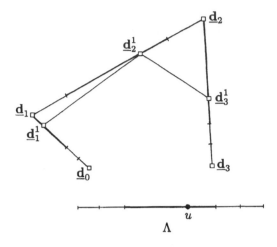

Figure 9.2. Inserting a knot: an example for $n = 3$.

9.2   The de Boor algorithm

Having inserted a new knot u into the sequence once, we may repeat the process and insert it again. We will obtain a new polygon $\mathbf{P}^{\lambda}[u, u]$:

$$\mathbf{P}^{\lambda}[u, u] = \mathbf{d}_0^{\lambda}\mathbf{d}_1^1(u)\mathbf{d}_2^2(u)\dots\mathbf{d}_n^2(u)\mathbf{d}_n^1(u)\mathbf{d}_n^{\lambda},$$

where the new $\mathbf{d}_i^2(u)$ are obtained by projective maps Φ_i^{n-1}

$$\underline{\mathbf{d}}_i^2(u) = \Phi_i^{n-1}(u); \quad i = 2, \dots, n,$$

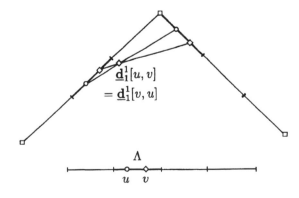

Figure 9.3. Knot insertion: two knots may be inserted in either order.

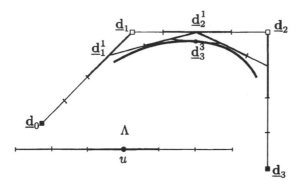

Figure 9.4. The de Boor algorithm: an example.

with

$$\Phi_i^{n-1} : \lambda_i^{n-1} \to \underline{\mathbf{d}}_{i-1}^1 \underline{\mathbf{d}}_i^1; \quad i = 2, \dots, n.$$

Here, the λ_i^{n-1} are spans of length $n-1$ that contain λ.

We may repeat this process until we have inserted u a total of n times, yielding a point $\underline{\mathbf{d}}_n^n(u)$ defined by

$$\underline{\mathbf{d}}_n^n(u) = \Phi_n^1(u).$$

We say that $\underline{\mathbf{d}}_n^n(u)$ is a point on a B-spline curve and call the above procedure the *de Boor algorithm* for its evaluation, see [39]. Thus,

$$\underline{\mathbf{d}}_i^r(u) = \Phi_i^{n-r+1}(u); \quad r = 1, \dots, n; \quad i = r, \dots, n. \tag{9.2}$$

Here, we have defined the projective maps $\Phi_i^{n-r+1}(u)$ by

$$\Phi_i^{n-r+1}(u) : \lambda_i^{n-r+1} \to \mathbf{d}_{i-1}^r \mathbf{d}_i^r.$$

Since projective maps in $I\!\!P^2$ correspond to linear interpolation, we may also write (9.2) as

$$\underline{\mathbf{d}}_i^r(u) = \frac{u_{i+n-r} - u}{u_{i+n-r} - u_{i-1}} \underline{\mathbf{d}}_{i-1}^{r-1}(u) + \frac{u - u_{i-1}}{u_{i+n-r} - u_{i-1}} \underline{\mathbf{d}}_i^{r-1}(u), \tag{9.3}$$

where we are now using the global numbering of control points.

Fig. 9.4 gives an example; Example 9.1 shows the algebra. Note that all third coordinates are unity: The curve is piecewise polynomial in affine space.

$$\begin{bmatrix} 1 \\ 1 \\ 1 \end{bmatrix}$$

$$\begin{bmatrix} 3 \\ 3 \\ 1 \end{bmatrix} \begin{bmatrix} 2.50 \\ 2.50 \\ 1.00 \end{bmatrix}$$

$$\begin{bmatrix} 6 \\ 3 \\ 1 \end{bmatrix} \begin{bmatrix} 4.50 \\ 3.00 \\ 1.00 \end{bmatrix} \begin{bmatrix} 3.83 \\ 2.83 \\ 1.00 \end{bmatrix}$$

$$\begin{bmatrix} 6 \\ 0 \\ 1.00 \end{bmatrix} \begin{bmatrix} 6.0 \\ 2.25 \\ 1.00 \end{bmatrix} \begin{bmatrix} 5.00 \\ 2.75 \\ 1.00 \end{bmatrix} \begin{bmatrix} 4.42 \\ 2.79 \\ 1.00 \end{bmatrix}$$

Example 9.1. The de Boor algorithm: the leftmost column contains the relevant control points. The relevant part of the knot sequence is 10, 11, 12, 14, 15, 16; we selected $\lambda = [12, 14]$ and $u = 13$.

Fig. 9.5 contains the control polygon from Fig. 9.4 as a subpolygon; the curve segment form Fig. 9.4 is highlighted in gray. The curve is evaluated for all parameter values in the domain interval $[8, 17]$.

What about NURBS in affine space? There, they become rational curves, just as was the case with Bézier curves. All we have to do is to project the final result $\underline{\mathbf{d}}_n^n(u)$ into affine space.

Figure 9.5. NURBS: an example curve. Knot sequence: 8, 8, 8, 10, 11, 12, 14, 15, 16, 17, 17, 17. All weights are unity.

Figure 9.6. NURBS: a B-spline curve in the projective plane with all of its individual segments shown as complete curves.

9.3 Properties

In the previous section, we already hinted that B-spline curves are piecewise polynomial[2]. Let us now verify that this is indeed so. Suppose, without any loss of generality, that a knot sequence contains the interval $\lambda = [0, 1]$. Suppose next that we insert both 0 and 1 into the knot sequence until they each have multiplicity n. Now, the de Boor algorithm (9.3) reduces to

$$\underline{\mathbf{d}}_i^r(u) = (1 - u)\underline{\mathbf{d}}_{i-1}^{r-1} + u\underline{\mathbf{d}}_i^{r-1},$$

which is nothing but the de Casteljau algorithm—except for a different subscripting scheme. Since the insertion of the knots 0 and 1 does not influence subsequent evaluation at u, we have shown that the de Boor algorithm yields a polynomial curve if we evaluate for $u \in \lambda$. Similar arguments hold for all remaining curve segments, and thus our curve is *piecewise polynomial*.

Fig. 9.6 shows a B-spline curve, but here we have drawn all polynomial segments in their entirety.[3] You see that all polynomial curves touch the horizon, indicating that they correspond to polynomial curves in the affine plane.

A similar argument may be used to determine how smooth a B-spline curve is, i.e., how often it is continuously differentiable. Take two neighbor-

---

[2]Recall we are in $I\!\!P^2$!

[3]This figure was inspired by Fig. X.2 from de Boor's "Practical Guide" [40].

Figure 9.7. NURBS: a B-spline curve in the projective plane with two osculants at the junction points shown as complete curves.

ing intervals $[u_{i-1}, u_i]$ and $[u_i, u_{I+1}]$. Assume the knot u_i that they both have in common has multiplicity k. Let us now insert u_{i-1} until this knot has multiplicity n, and then repeat the same process for u_{i+1}. If we now evaluate at u, we do nothing but perform $n - k$ steps of the de Casteljau algorithm relative to the interval $[u_{i-1}, u_{i+1}]$. Because of the subdivision property of Bézier curves, we see that the resulting curve is C^{n-k} at u_i. Therefore, if a knot has multiplicity k, then the curve is C^{n-k} at that knot. It may in fact be of higher-order smoothness, but it has at least $n - k$ continuous derivatives.

Fig. 9.7 illustrates how a cubic B-spline curve with simple knots is a C^2 curve. Recall that a cubic Bézier curve has an *osculant* at each endpoint — these are conics that agree with the cubic in position, first and second derivative. If two cubics are C^2, then they have a common osculant. The composite curve from Fig. 9.7 consists of three cubic segments; the osculants at the two junction points are shown; they are parabolas, so the B-spline curve is polynomial in affine space.

Typically, the end knots are of multiplicity n: $u_0 = \ldots = u_{n-1}$ and $u_{K-n+1} = \ldots = u_K$. In this case, the first and last control points, $\underline{\mathbf{d}}_0$ and $\underline{\mathbf{d}}_p$, are on the curve. This is the most desirable configuration for the end knots, as it allows more direct control over the curve shape. In the IGES data specification (Section 15.1), the first and last knots would have to be specified $n + 1$ times each — not necessary, but now common usage by many systems. If the end knots are not of multiplicity n, the first and last

Figure 9.8. NURB weights: top, an integral NURB curve; bottom, after two weights were set to 0.1.

control points are not on the curve and are called "phantom vertices" [16].

For Bézier curves in $I\!\!P^2$, we saw how multiplying a control point's coordinates by a constant could change the shape of the curve. The same happens here; again calling the last coordinate w_i a *weight*, Fig. 9.8 illustrates what happens if the weight is changed.[4] Each point on the curve is associated with a parameter value u. We can also associate each control point $\underline{\mathbf{d}}_i$ with a parameter value ξ_i, defined as

$$\xi_i = \frac{1}{n}(u_i + \ldots + u_{i+n-1}).$$

The ξ_i are called *Greville abscissae*, see [57].

We may again introduce *weight points* $\mathbf{q}_i$ by setting

$$\mathbf{q}_i = \frac{w_i \mathbf{d}_i + w_{i+1}\mathbf{d}_{i+1}}{w_i + w_{i+1}}.$$

Each weight point is associated with the parameter value $\zeta_i = \frac{1}{2}\xi_i + \frac{1}{2}\xi_{i+1}$.

These weight points may be used for shape control in exactly the same way as in the case of rational Bézier curves. If some of the weights are negative, the weight points will not be between their corresponding control points. In these cases, singularities are possible.

---

[4] Changing w_i to αw_i means, of course, that *all* coordinates of $\underline{\mathbf{d}}_i$ are multiplied by α!

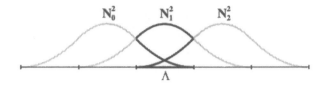

Figure 9.9. B-splines: three quadratic B-splines N_i^2 that do not vanish over an interval λ.

We finally note one of the most important properties of B-spline curves: *local control*. If we move one $\mathbf{d}_k$, the curve will change, but only "near" $\mathbf{d}_k$. This follows since the process of knot insertion is itself local. And, moreover, if $\mathbf{d}_k$ is changed, then the curve changes only for those u with $u \in [u_{k-1}, u_{k+n}]$.

9.4 The B-spline Form

In the case of Bézier curves, we derived the explicit Bernstein form from the de Casteljau algorithm. The analogous reasoning for B-splines yields the following explicit form:

$$\underline{s}(u) = \sum_{i=0}^{p} \mathbf{d}_i N_i^n(u); \quad \mathbf{d}_i \in \mathbb{P}^3, \tag{9.4}$$

assuming as before that $u \in \lambda$ for some interval $\lambda = [u_I, u_{I+1}]$. The functions N_i^n are piecewise polynomial. Since B-spline curves enjoy the local control property, the corresponding basis functions have *local support*: Each N_i^n is nonzero only over the interval $[u_{i-1}, u_{i+n}]$. Fig. 9.9 shows examples.

The basis functions N_i^n satisfy the following recursion:

$$N_l^n(u) = \frac{u - u_{l-1}}{u_{l+n-1} - u_{l-1}} N_l^{n-1}(u) + \frac{u_{l+n} - u}{u_{l+n} - u_l} N_{l+1}^{n-1}(u). \tag{9.5}$$

The recursion is anchored by

$$N_i^0(u) = \begin{cases} 1 & \text{if } u_{i-1} \leq u < u_i \\ 0 & \text{else} \end{cases}. \tag{9.6}$$

While the existence of the explicit form (9.4) is of theoretical importance, we should point out that evaluation of a curve is much more efficient using the de Boor algorithm!

In affine space, the explicit form corresponds to the *rational B-spline* formulation:

$$\underline{s}(u) = \frac{\sum_{i=0}^{p} w_i \mathbf{d}_i N_i^n(u)}{\sum_{i=0}^{p} w_i N_i^n(u)}; \quad \underline{\mathbf{d}}_i \in I\!\!P^3. \tag{9.7}$$

We may also write this as

$$\underline{s}(u) = \sum_{i=0}^{p} \mathbf{d}_i R_i^n(u) \tag{9.8}$$

with appropriately defined functions R_i^n:

$$R_i^n(u) = \frac{w_i N_i^n(u)}{\sum_{j=0}^{p} w_j N_j^n(u)}.$$

9.5   Blossoms

Blossoms are a nice tool for reasoning about Bézier curves; for B-splines, they turn out to be an indispensable practical tool as well. We define a blossom via the de Boor algorithm: after fixing an interval λ, instead of feeding in one parameter value u into the algorithm at every level, we use different parameters u_i for each level. We then arrive, in complete analogy to blossoms of Bézier curves, at the B-spline blossom $\underline{\mathbf{d}}[v_1, \ldots, v_n]$.

The de Boor algorithm for an interval λ, when generalized to the blossom form, will look like this (cubic case):

$$
\begin{array}{llll}
\underline{\mathbf{d}}_0 & & & \\
\underline{\mathbf{d}}_1 & \underline{\mathbf{d}}_1^1[v_1] & & \\
\underline{\mathbf{d}}_2 & \underline{\mathbf{d}}_2^1[v_1] & \underline{\mathbf{d}}_2^2[v_1, v_2] & \\
\underline{\mathbf{d}}_3 & \underline{\mathbf{d}}_3^1[v_1] & \underline{\mathbf{d}}_3^2[v_1, v_2] & \underline{\mathbf{d}}_3^3[v_1, v_2, v_3].
\end{array}
$$

As for the Bézier case, we will not write $\underline{\mathbf{d}}_3^3[v_1, v_2, v_3]$ for the final result, but simply $\underline{\mathbf{d}}^\lambda[v_1, v_2, v_3]$.

Some experimentation with blossoms, just as for the Bézier case, shows that the curve segment defined over $\lambda = [u_I, u_{I+1}]$ needs $n+1$ control points for its definition, namely $\mathbf{d}_{I-n+1}, \ldots, \mathbf{d}_{I+1}$. They can be obtained as blossom values:

$$\mathbf{d}_{I-n+1+k} = \mathbf{d}^\lambda[u_{I-n+1+k}, \ldots, u_{I+k}]; \quad k = 0, \ldots, n. \tag{9.9}$$

The arguments of $\mathbf{d}^\lambda$ on the right-hand side are all n-tuples of subsequent knots that contain either u_I or u_{I+1}.

Figure 9.10. NURBS: a B-spline curve with the B-spline control polygon (white squares) as well as the piecewise Bézier form (gray squares). Degree: three.

If we are interested in the Bézier points $\underline{\mathbf{b}}_i$ of the curve segment corresponding to λ, all we have to do is to evaluate the blossom

$$\underline{\mathbf{b}}_i = \underline{\mathbf{d}}^\lambda[u_I < n - i >, u_{I+1}^{<i>}]; \quad i = 0, \ldots, n.$$

As before, l and r are the two successive and distinct knots that define λ. Fig. 9.10 shows a B-spline curve with its original B-spline polygon and also with the Bézier polygon of each piece; both are representations of the same curve! The knot sequence for the original B-spline curve is 0, 0, 0, 1, 2, 3, 4, 4, 4. The knot sequence for the piecewise Bézier curve is simply 0, 0, 0, 1, 1, 1, 2, 2, 2, 3, 3, 3, 4, 4, 4.

9.6   Derivatives

Integral B-spline curves of degree n are $n - r$ times differentiable at a knot that has multiplicity r. At locations other than the knots, they are infinitely often differentiable, since there they behave like integral polynomials. The rational case is more complicated, of course: Derivatives may not exist at singularities, and they become very involved for higher derivatives.

While for higher derivatives the recursive formula (7.31) works,[5] we are luckier for the first derivative.

The de Boor algorithm for the homogeneous points $\underline{\mathbf{d}}_i$ yields the curves' tangent: It is spanned by $\underline{\mathbf{d}}_{n-1}^{n-1} \wedge \underline{\mathbf{d}}_n^{n-1}$. The derivative of the projective curve is given by

$$\underline{\dot{\mathbf{s}}}(u) = \frac{n}{u_{I+1} - u_I} \left[\underline{\mathbf{d}}_n^{n-1} - \underline{\mathbf{d}}_{n-1}^{n-1} \right].$$

---

[5] We just have to interpret the involved $\mathbf{p}(t), w(t), \mathbf{b}(t)$ as B-splines.

Unfortunately, it does not project to the derivative of the rational curve. But following a result by Floater [71], we have

$$\dot{\mathbf{s}}(u) = \frac{n}{u_{I+1} - u_I} \frac{w_{n-1}^{n-1} w_n^{n-1}}{[w_n^n]^2} \left[\mathbf{d}_n^{n-1} - \mathbf{d}_{n-1}^{n-1} \right]. \tag{9.10}$$

This may be used to obtain a *bound* on the derivatives of a NURB curve, quite in analogy to (7.34):

$$\| \dot{\mathbf{s}}(u) \| \leq n \left[\frac{\max\{w_i\}}{\min\{w_i\}} \right]^2 \max_{0 < i \leq p-1} \left\{ \frac{\| \mathbf{d}_i - \mathbf{d}_{i-1} \|}{u_{i+n-1} - u_{i-1}} \right\}. \tag{9.11}$$

Note that this is a global estimate; a localization to an interval λ does not make sense here.

9.7   More NURB properties

Rational B-splines have properties that should not be surprising at this point; we list the most important ones:
Projective invariance: The following two operations yield the same result.

a) evaluate a NURB curve for a parameter value u, yielding a point $\mathbf{s}(u)$, then transform $\mathbf{s}(u)$ by a projective map to a point $\Phi\mathbf{s}(u)$.

b) transform the control polygon plus weights by Φ, then evaluate at u.

Note that step b) will change the weights. In case the projective transformation was actually affine, we retain affine invariance: The weights will not change.
Convex hull property: If the weights are nonnegative, the curve lies in the convex hull of the control polygon. In fact, any point on the curve lies in the convex of just $n + 1$ consecutive control points.
Variation diminishing property: This holds provided the weights are nonnegative.
Superset property: NURBS are a true superset of rational Bézier curves. In order to write a rational Bézier curve of degree n as a NURB curve, define a knot sequence

$$\{u_0, u_1, \ldots, u_{2n-1}\} = \{0^{<n>}, 1^{<n>}\}.$$

Note that the Greville abscissae now satisfy $\xi_i = i/n$.
    We may also **reparametrize** NURBS in the same sense as we could rational Bézier curves. Let a Moebius transformation map the knot sequence $\{u_i\}$ to a new knot sequence $\{\hat{u}_i\}$, such that the first and last knots

Figure 9.11. Reparametrization: top, the initial curve; bottom, after reparametrization. Squares denote B-spline control vertices, circles denote Bézier control vertices, and diamonds denote weight points.

are not changed. The weight point parameter values ζ_i will be mapped to new values $\hat{\zeta}_i$. The weight points themselves will be mapped to new weight points $\hat{\mathbf{q}}_i$, and the following identity must hold:

$$\mathrm{cr}(\mathbf{d}_i, \mathbf{q}_i, \hat{\mathbf{q}}_i, \mathbf{d}_{i+1}) = \mathrm{cr}(\xi_i, \zeta_i, \hat{\zeta}_i, \xi_{i+1}). \tag{9.12}$$

This is in analogy to the Bézier case (7.22), but now things are more complicated: In (7.22), the right-hand side was the same for all i, whereas now it depends on i! NURB reparametrization was first addressed by E. Lee and M. Lucian [100].[6] Fig. 9.11 illustrates.

B-spline curves in projective space or NURB curves in affine space may also be **closed**. Then the knot sequence is *periodic* by simply counting it mod(K). An illustration of a closed NURB curve together with its piecewise Bézier polygon is given in Fig. 9.12.

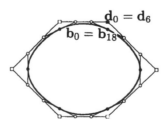

Figure 9.12. Closed NURBS: a cubic example.

---

[6]That article corrects a false statement made by myself in [53].

9.8   Problems

1. Find a recursion, similar to (9.5) for the functions R_j^n.

2. Find a formula similar to (7.33) for NURBS. How does it compare to (9.11)?

3. Is a NURB curve in the convex hull of $d_0, q_0, q_1, \ldots$, in analogy to the Bézier case as illustrated in Fig. 7.5?

10

Pythagorean Curves

In this chapter, we will be firmly rooted in Euclidean geometry. The curves
we are going to discuss all satisfy a property that makes some associated
vectors be of unit length, for example, tangent vectors.

10.1   Circles and the stereographic projection

A well-known parametrization of the circle may be obtained as follows.
Consider a unit circle with center at the origin of a Euclidean coordinate
system. We can write the southeast quarter as a rational quadratic with
control points

$$\begin{bmatrix} 0 \\ -1 \end{bmatrix}, \begin{bmatrix} 1 \\ -1 \end{bmatrix}, \begin{bmatrix} 1 \\ 0 \end{bmatrix}$$

and standard weights $1, \frac{1}{2}\sqrt{2}, 1$. We may also reparametrize this circle
representation, using a reparametrization constant c.[1] The equation then
becomes

$$\begin{bmatrix} x \\ y \end{bmatrix} = \begin{bmatrix} \frac{c\sqrt{2}(1-t)t+c^2t^2}{(1-t)^2+c\sqrt{2}(1-t)t+c^2t^2} \\ \frac{-(1-t)^2-c\sqrt{2}(1-t)t}{(1-t)^2+c\sqrt{2}(1-t)t+c^2t^2} \end{bmatrix}, \tag{10.1}$$

---

[1]We are referring to a rational linear reparametrization as outlined in Section 4.9.

169

with $c = 1$ for the familiar standard form. Let us try to arrange the parametrization such that the curve passes through $\mathbf{n} = [0, 1]^T$, the north pole of the circle, for $t = \infty$. This leads to

$$
\begin{bmatrix} 0 \\ 1 \end{bmatrix} = \begin{bmatrix} \frac{c^2 - c\sqrt{2}}{c^2 - c\sqrt{2} + 1} \\ \frac{c\sqrt{2} - 1}{c^2 - c\sqrt{2} + 1} \end{bmatrix},
$$

from which we deduce $c = \sqrt{2}$. We may then take the reparametrized Bézier form (10.1) and rewrite it as follows:

$$
\begin{bmatrix} x(t) \\ y(t) \end{bmatrix} = \begin{bmatrix} \frac{2t}{t^2 + 1} \\ \frac{t^2 - 1}{t^2 + 1} \end{bmatrix}. \tag{10.2}
$$

Check that $x^2(t) + y^2(t) = 1$!

This parametrization is an example of the *stereographic projection*: the line through the north pole $\mathbf{n}$ and $[x(t), y(t)]^T$ intersects the x-axis in $x = t$. This is seen by checking that the three points

$$
\begin{bmatrix} 0 \\ 1 \end{bmatrix}, \quad \begin{bmatrix} \frac{2t}{t^2 + 1} \\ \frac{t^2 - 1}{t^2 + 1} \end{bmatrix}, \quad \begin{bmatrix} t \\ 0 \end{bmatrix}
$$

are collinear.

Thus every point on the x-axis has a unique image point on the circle and vice versa. Fig. 10.1 illustrates this. More on these maps is presented in Section 13.3.

The stereographic projection may be used to obtain *all* rational parametrizations of the circle, see [124]. If we replace t by a rational function $f(t)/g(t)$ (both f and g being polynomials), then the corresponding point on the circle is given by

$$
\begin{bmatrix} x(t) \\ y(t) \end{bmatrix} = \begin{bmatrix} \frac{2[\frac{f(t)}{g(t)}]}{[\frac{f(t)}{g(t)}]^2 + 1} \\ \frac{[\frac{f(t)}{g(t)}]^2 - 1}{[\frac{f(t)}{g(t)}]^2 + 1} \end{bmatrix},
$$

which can be simplified to

$$
\begin{bmatrix} x(t) \\ y(t) \end{bmatrix} = \begin{bmatrix} \frac{2f(t)g(t)}{f^2(t) + g^2(t)} \\ \frac{f^2(t) - g^2(t)}{f^2(t) + g^2(t)} \end{bmatrix}. \tag{10.3}
$$

If we define polynomials $a(t), b(t), c(t)$ by setting

$$
a(t) = 2f(t)g(t), \quad b(t) = f^2(t) - g^2(t), \quad c(t) = f^2(t) + g^2(t), \tag{10.4}
$$

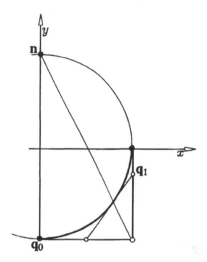

Figure 10.1. The circle: a parametrization that corresponds to the stereographic projection.

then our point on the circle may be written as

$$\begin{bmatrix} x(t) \\ y(t) \end{bmatrix} = \begin{bmatrix} \frac{a(t)}{c(t)} \\ \frac{b(t)}{c(t)} \end{bmatrix}.$$

Since a, b, and c are of the form (10.4), it is clear that we have

$$a^2(t) + b^2(t) = c^2(t). \tag{10.5}$$

This equation is reminiscent of Pythagoras' theorem; we thus call the polynomials a, b, and c *Pythagorean triples*.

Thus for any choice of polynomials f and g, there is a corresponding (even degree) parametrization of the form (10.3). That in fact *all* rational parametrizations of the circle are of this form goes back to a theorem by Kubota [97], which states that any three polynomials satisfying (10.5) must be of the form (10.4). Three different parametrizations of a semicircle are shown in Fig. 10.2. The middle one corresponds to a constant speed parametrization, also referred to as the *arc length parametrization*.

When we represent a circle by a rational Bézier curve such as (10.2), the entire circle is covered once by the Bézier curve. The region corresponding to the point $t = \pm\infty$ is hard to "reach". This may be avoided: It is possible to write a full circle as one rational Bézier curve of degree five, see

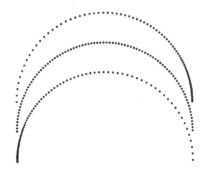

Figure 10.2. Circle parametrizations: three different parametrizations of a semicircle.

Chou [35]. Its Bézier points are given by

$$\left[\begin{bmatrix} 1 \\ 0 \end{bmatrix} \begin{bmatrix} 1 \\ 4 \end{bmatrix} \begin{bmatrix} -3 \\ 2 \end{bmatrix} \begin{bmatrix} -3 \\ -2 \end{bmatrix} \begin{bmatrix} 1 \\ -4 \end{bmatrix} \begin{bmatrix} 1 \\ 0 \end{bmatrix} \right],$$

and its weights are

$$[1, \frac{1}{5}, \frac{1}{5}, \frac{1}{5}, \frac{1}{5}, 1].$$

As the parameter t traces out all values between $-\infty$ and $+\infty$, the circle is traced out twice. Such curves are *improperly parametrized*; see also Section 7.7.

Why would anyone be interested in different parametrizations of a circle? Recall that the parametrization corresponds to the *speed* with which the circle is traversed. If, for instance, the circle describes how a robot moves an object, then the speed which it "delivers" the object may well be of critical importance![2]

10.2   Pythagorean hodograph curves

Let $\mathbf{x}(t)$ denote an arbitrary planar parametric curve. Consider the part of the curve between $t = 0$ and some arbitrary $t = r$. The length s of this

---

[2] An interesting industrial application was told to me by J. Hoschek: a robot arm had to grab labels and then paste them on beer bottles. If the "delivery speed" was too high, the bottles broke — so a careful analysis of the parametrization of the robot arm path was needed in order to optimize the labeling process.

curve segment is given by

$$s = s(r) = \int_0^r \|\dot{\mathbf{x}}(t)\| dt, \qquad (10.6)$$

as may be found in any text on differential geometry, see [46] and [57].
The detailed expression for $\|\dot{\mathbf{x}}(t)\|$ is

$$\|\dot{\mathbf{x}}(t)\| = \sqrt{\dot{x}^2 + \dot{y}^2},$$

and because of the square root, $\|\dot{\mathbf{x}}(t)\|$ will in general not be a rational function of t. If it were, we could integrate (10.6) explicitly, without having to rely on numerical quadrature.

Farouki and Sakkalis [63] had the idea to investigate curves for which $\|\dot{\mathbf{x}}(t)\|$ is a rational function of t, and they started by investigating the polynomial case. For such curves, $\sqrt{\dot{x}^2 + \dot{y}^2}$ would itself be a polynomial, say $p(t)$. Squaring this gives

$$\dot{x}(t)^2 + \dot{y}(t)^2 = p^2(t). \qquad (10.7)$$

This equation is of the form (10.4) from Section 10.1. From that section, we know that we can generate curves satisfying (10.7) simply by setting

$$\dot{x}(t) = 2f(t)g(t), \quad \dot{y}(t) = f^2(t) - g^2(t), \quad p(t) = f^2(t) + g^2(t) \qquad (10.8)$$

for some polynomials f and g. Such curves are known as *Pythagorean hodograph curves* — or PH curves for short — since their defining property (10.7) involves Pythagorean triples.

Let us explicitly construct an example. We choose (arbitrarily)

$$
\begin{aligned}
f(t) &= (1-t)f_0 + tf_1, \\
g(t) &= (1-t)g_0 + tg_1.
\end{aligned}
$$

We then compute $\dot{x}$ and $\dot{y}$:

$$
\begin{aligned}
\dot{x}(t) &= 2[(1-t)f_0 + tf_1][(1-t)g_0 + tg_1], \\
\dot{y}(t) &= [(1-t)f_0 + tf_1]^2 - [(1-t)g_0 + tg_1]^2.
\end{aligned}
$$

This derivative — or hodograph — is guaranteed to satisfy (10.7). In order to obtain the actual curve $\mathbf{x}(t)$, we simply integrate, component by component, using an integration constant $\mathbf{b}_0$. Integrating quadratic polynomials yields cubic polynomials, resulting in a cubic curve $\mathbf{x}(t)$. We can write it

Figure 10.3. Pythagorean hodograph curves: two instances of Tschirnhausen's cubic in $\mathbb{P}^2$.

in Bézier form, following Farouki [63]: The first Bézier point is given by $\mathbf{b}_0$; the remaining ones are

$$\mathbf{b}_1 = \mathbf{b}_0 + \frac{1}{3}\begin{bmatrix} f_0^2 - g_0^2 \\ 2f_0g_0 \end{bmatrix}$$

$$\mathbf{b}_2 = \mathbf{b}_1 + \frac{1}{3}\begin{bmatrix} f_0f_1 - g_0g_1 \\ f_0g_1 + f_1g_0 \end{bmatrix}$$

$$\mathbf{b}_3 = \mathbf{b}_2 + \frac{1}{3}\begin{bmatrix} f_1^2 - g_1^2 \\ 2f_1g_1 \end{bmatrix}.$$

This kind of Bézier curve does not have much intuitive appeal — the free parameters f_0, f_1, g_0, g_1 do not have any geometric meaning. But there is a geometric interpretation; Cubic PH curves in Bézier form are uniquely characterized by two conditions:

$$\|\Delta\mathbf{b}_1\|^2 = \|\Delta\mathbf{b}_0\|\|\Delta\mathbf{b}_2\| \quad \text{and} \quad \angle(\mathbf{b}_0, \mathbf{b}_1, \mathbf{b}_2) = \angle(\mathbf{b}_1, \mathbf{b}_2, \mathbf{b}_3).$$

While this gives some nice geometric insight, it also reveals that cubic PH curves cannot offer much for the design of planar free-form curves: they are simply too restricted.

Further analysis reveals that the above two conditions only characterize one cubic, except for scalings or rigid body motions. It is called *Tschirnhausen's cubic*; Fig. 10.3 shows two instances in the extended affine plane.

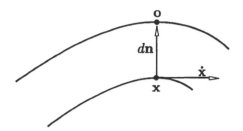

Figure 10.4. Offset curves: illustration of the geometry.

PH curves are not restricted to cubics. Different choices of f and g above can produce polynomials of arbitrary degree. Farouki [61] constructed quintic PH splines and has used them successfully to control step lengths in numerically controlled milling [65], [66], [62]. Walton and Meek used quintic PH splines for the approximation of spirals, see [152] and [17].

10.3 Rational PH offsets

One of the standard problems in CAD/CAM is that of computing *offsets*: planar curves that are at a fixed distance from a given curve. If x is a given curve, then a curve o with distance d to x is defined by

$$o(t) = x(t) + dn(t), \qquad (10.9)$$

where

$$n(t) = \frac{1}{\sqrt{[\dot{x}(t)]^2 + [\dot{y}(t)]^2}} \begin{bmatrix} \dot{y}(t) \\ -\dot{x}(t) \end{bmatrix} \qquad (10.10)$$

is the unit normal of the curve. See Fig. 10.4 for an illustration of the defining terms and Fig. 10.5 for an example.

Due to the appearance of the square root in (10.9), the offset to a rational curve will in general not be a rational curve itself. We know one exception already, however: PH curves! For them, (10.7) holds, so those curves do have rational offsets. For the example of a cubic PH curve, the offset will be rational quintic: In (10.9), the involved terms are of the form

$$o(t) = \mathbf{cubic}(t) + d\frac{\mathbf{quadratic}(t)}{\mathbf{quadratic}(t)};$$

this becomes rational quintic after finding a common denominator.

Figure 10.5. Offset curves: several offsets (gray) in both directions of the center curve (black).

The homogeneous control points $\underline{o}_i$ of the offset curve are given by

$$
\begin{aligned}
\underline{o}_0 &= \sigma_0\underline{b}_0 + 3d[\Delta\underline{b}_0]^{\perp}, \\
\underline{o}_1 &= \frac{1}{5}\left[2\sigma_1\underline{b}_0 + 3\sigma_0\underline{b}_1 + 3d[3\Delta\underline{b}_0 + 2\Delta\underline{b}_1]^{\perp}\right], \\
\underline{o}_2 &= \frac{1}{10}\left[\sigma_2\underline{b}_0 + 6\sigma_1\underline{b}_1 + 3\sigma_0\underline{b}_2 + 3d[3\Delta\underline{b}_0 + 6\Delta\underline{b}_1 + \Delta\underline{b}_2]^{\perp}\right], \\
\underline{o}_3 &= \frac{1}{10}\left[3\sigma_2\underline{b}_1 + 6\sigma_1\underline{b}_2 + \sigma_0\underline{b}_3 + 3d[\Delta\underline{b}_0 + 6\Delta\underline{b}_1 + 3\Delta\underline{b}_2]^{\perp}\right], \\
\underline{o}_4 &= \frac{1}{5}\left[3\sigma_2\underline{b}_2 + 2\sigma_0\underline{b}_3 + 3d[2\Delta\underline{b}_1 + 3\Delta\underline{b}_2]^{\perp}\right], \\
\underline{o}_5 &= \sigma_2\underline{b}_3 + 3d[\Delta\underline{b}_2]^{\perp}, \tag{10.11}
\end{aligned}
$$

where

$$
\begin{aligned}
\sigma_0 &= f_0^2 + g_0^2, \\
\sigma_1 &= f_0f_1 + g_0g_1, \\
\sigma_2 &= f_1^2 + g_1^2.
\end{aligned}
$$

The $\perp$ symbol denotes the Euclidan perpendicular vector:

$$
\begin{bmatrix} x \\ y \\ 0 \end{bmatrix}^{\perp} = \begin{bmatrix} y \\ -x \\ 0 \end{bmatrix}.
$$

In (10.11), all you have to do is vary d and you get any desired offset to the original cubic PH curve!

10.4 Rational offsets

PH curves are very special polynomial curves. Since their offsets are rational, one might ask if there are *rational* curves with rational offsets. The definition of such curves is still given by (10.9) except that now $\mathbf{x}$ is a rational curve. Clearly, the condition for a rational offset is that $\mathbf{n}$, the unit normal vector, is a rational function of t.

If we plot all vectors $\mathbf{n}(t)$ emanating from the origin, then their locus will be on the unit circle. In other words, we are looking for rational parametrizations of the unit circle — just as we discussed at the end of Section 10.1! Thus if the normal vector $\mathbf{n}(t)$ can be written as

$$\mathbf{n}(t) = \begin{bmatrix} n_x(t) \\ n_y(t) \end{bmatrix} = \begin{bmatrix} \frac{m(t)}{v(t)} \\ \frac{n(t)}{v(t)} \end{bmatrix},$$

then the three polynomials $m(t), n(t)$, and $v(t)$ must be of the form

$$m(t) = 2f(t)g(t), \quad n(t) = f^2(t) - g^2(t), \quad v(t) = f^2(t) + g^2(t), \quad (10.12)$$

for the same reason that led to (10.4).

Thus

$$\mathbf{n}(t) = \begin{bmatrix} \frac{2f(t)g(t)}{f^2(t)+g^2(t)} \\ \frac{f^2(t)-g^2(t)}{f^2(t)+g^2(t)} \end{bmatrix} \tag{10.13}$$

for some polynomials $f(t)$ and $g(t)$.

We now follow an argument by Pottmann [124] to arrive at a characterization of the rational curve itself, not just of its normal. At a point $\mathbf{x}(t)$ of the curve, the tangent is perpendicular to the normal and has an implicit equation of the form $ax + by = c$. Since it is also a rational function of t, we may write this implicit equation as

$$n_x(t)x + n_y(t)y = r(t)) \tag{10.14}$$

with rational function $r(t)$. Here, x and y are generic points, not just those on the curve. Those x and y that satisfy (10.14) are on the tangent at parameter value t. The value of $r(t)$ then denotes the distance of the tangent from the origin.

We may differentiate (10.14) with respect to t:

$$\dot{n}_x(t)x + \dot{n}_y(t)y = \dot{r}(t). \tag{10.15}$$

Now (10.14) and (10.15) constitute a 2×2 linear system for x and y, with the solution

$$x \;=\; \frac{2fg}{f^2 + g^2}r - \frac{f^2 - g^2}{2(\dot{f}g - f\dot{g})}\dot{r}, \tag{10.16}$$

$$y \;=\; \frac{f^2 - g^2}{f^2 + g^2}r - \frac{fg}{\dot{f}g - f\dot{g}}\dot{r}. \tag{10.17}$$

Since $r(t)$ is the distance from the tangent to the origin, replacing $r(t)$ above by $r(t) + d$ yields the offset of distance d to the curve $\mathbf{x}(t)$.

10.5   Rational curves on the sphere

In Section 10.1 we gave a characterization of all rational parametrizations of a circle. A more general question is to ask for a characterization of all rational curves whose locus is on a sphere. Without any loss of generality, we will use the unit sphere. As in the circle case, it will be useful to employ the concept of the *stereographic projection*.

In the 2D case, a curve that "lives" on the unit circle has to satisfy (10.4). The 3D case then considers curves that "live" on the unit circle. Similar conditions hold, and their proof is mathematically involved (Dietz *et al.* [45]).

Any point $\underline{\mathbf{x}}$ on the unit sphere[3] satisfies

$$x^2 + y^2 + z^2 = w^2.$$

If $\underline{\mathbf{x}}$ traces out a polynomial[4] curve $\underline{\mathbf{x}}(t)$, then we must have

$$x(t)^2 + y(t)^2 + z(t)^2 = w(t)^2. \tag{10.18}$$

If four polynomials $x(t), y(t), z(t), w(t)$ satisfy (10.18), then they must be of the following form:

$$
\begin{aligned}
x &= \; p_0^2 + p_1^2 + p_2^2 + p_3^2 \;\; \text{or} \;\; x = -p_0^2 - p_1^2 - p_2^2 - p_3^2, \\
y &= \; 2p_0p_1 - 2p_2p_3, \\
z &= \; 2p_2p_3 + 2p_0p_2, \\
w &= \; p_1^2 + p_2^2 - p_0^2 - p_3^2,
\end{aligned}
\tag{10.19}
$$

where $p_i = p_i(t)$ are polynomials.

We see immediately that any (homogeneous) polynomial curve on the unit sphere must be of even degree! An example is shown in Fig. 10.6.

---

[3] We are working in extended affine 3-space here.

[4] Keep in mind that the polynomial curves are in homogeneous form — it is only after projecting into affine space that they actually are on a sphere!

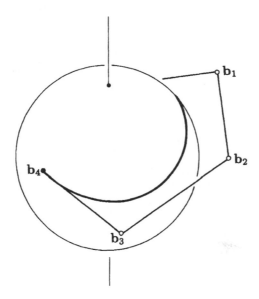

Figure 10.6. Curves on the sphere: a rational quartic curve together with its control points.

One particular way of obtaining curves on the unit sphere is by using the *stereographic projection*. It maps every point in the xy-plane to a point on the unit sphere by joining it to the "north pole" of the sphere and then finding the intersection of that line with the sphere. The pre-image point thus has three homogeneous coordinates while the image has four. The map is given by

$$
\begin{bmatrix} x \\ y \\ 0 \\ w \end{bmatrix} \rightarrow \begin{bmatrix} 2xw \\ 2yw \\ x^2 + y^2 - w^2 \\ x^2 + y^2 + w^2 \end{bmatrix}. \tag{10.20}
$$

As in the circle case, verify that the three points

$$
\begin{bmatrix} 0 \\ 0 \\ 1 \\ 1 \end{bmatrix}, \quad \begin{bmatrix} 2x \\ 2y \\ x^2 + y^2 - w^2 \\ x^2 + y^2 + w^2 \end{bmatrix}, \quad \begin{bmatrix} x \\ y \\ 0 \\ w \end{bmatrix}
$$

are collinear!

You will notice that the point $[0, 0, 1, 1]^T$, i.e., the north pole of the sphere, is not a unique map of a point in the plane. In fact, all 2D points $[x, y, 0, w]^T$ map to it. For all other points, the stereographic projection is one-to-one.

If $[x(t), y(t), 0, w(t)]^T$ is a Bézier curve in the plane, then (10.20) may be used to compute the Bézier points of its image under the stereographic projection. All that is needed is the identity

$$B_i^n(t) B_j^n(t) = \frac{\binom{n}{i}\binom{n}{j}}{\binom{2n}{i+j}} B_{i+j}^{2n}(t).$$

We will later revisit the stereographic projection — see Section 13.3.

10.6   Problems

1. In Section 10.1, we saw how to cover the whole circle by *twice* by a rational quintic. Is there a way to cover it *three* times?

2. Are all (even degree) curves on a sphere closed?

3. Define the term "inflection point" for a curve on a sphere. How many inflection points can a rational quartic on a sphere have?

11

Rectangular Patches

The theory of rational curves was an easy and straightforward extension of our treatment of conics. In the surface case, things will turn out to be more complicated. It would appear natural to base the development of surfaces on an extension of quadrics, but this is not possible. We therefore start with the parametric definition of a rational surface, shifting between affine and projective contexts as appropriate. We introduce rational Bézier patches and NURBS just as we did for the curve case. Again, the Bézier form is the basic building block, and we start from it.

11.1 Bilinear patches

A point $\mathbf{x}(u, v)$ on a rational bilinear patch is the image a point (u, v), defined by

$$\mathbf{x}(u, v) = \frac{\sum_{i=0}^{1} \sum_{j=0}^{1} w_{ij} \mathbf{b}_{ij} B_i^1(u) B_j^1(v)}{\sum_{i=0}^{1} \sum_{j=0}^{1} w_{ij} B_i^1(u) B_j^1(v)}. \tag{11.1}$$

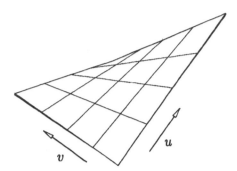

Figure 11.1. Bilinear patches: a patch is covered by two families of straight lines.

It may be viewed as a projection of bilinear polynomial patch in $\mathbb{P}^3$, given by

$$\underline{\mathbf{x}}(u,v) = \sum_{i=0}^{1} \sum_{j=0}^{1} \underline{\mathbf{b}}_{ij} B_i^1(u) B_j^1(v). \tag{11.2}$$

This patch is a map of a quadrilateral in the projective plane, defining a Moebius net as discussed in Section 1.5. The term *bilinear* is due to the fact that the patch equation is linear in u as well as in v. Geometrically, this means that the patch contains two families of straight lines: at a fixed point $\mathbf{x}(u^*, v^*)$, one line is given by

$$\underline{\mathbf{l}}(u) = (1-u)\big[(1-v^*)\underline{\mathbf{b}}_{00} + v^*\underline{\mathbf{b}}_{01}\big] + u\big[(1-v^*)\underline{\mathbf{b}}_{10} + v^*\underline{\mathbf{b}}_{11}\big],$$

the other one by

$$\underline{\mathbf{l}}(v) = (1-v)\big[(1-u^*)\underline{\mathbf{b}}_{00} + u^*\underline{\mathbf{b}}_{10}\big] + v\big[(1-u^*)\underline{\mathbf{b}}_{01} + u^*\underline{\mathbf{b}}_{11}\big].$$

These two lines span a plane, which is the *tangent plane* of our patch at (u^*, v^*).

Fig. 11.1 illustrates.[1] Since intersecting straight lines are a preserved under projections, the rational bilinear patch (11.1) also contains two families of straight lines. Such surfaces are called *doubly ruled*.

Let us now investigate some of the geometric properties of our patch, resorting to $\mathbb{P}^3$ for convenience. Let (u,v) trace out a straight line in the domain, i.e., we set $u = u(t)$ and $v = v(t)$, and both $u(t)$ and $v(t)$ are linear in t. Inserting this into (11.2), we obtain a curve $\underline{\mathbf{c}}(t) = \underline{\mathbf{x}}(u(t), v(t))$, lying completely on the patch. It does not take much effort to realize that $\mathbf{c}(t)$

---

[1]More bilinear patches are shown in Chapter 13; another example is utilized in Fig. 8.5.

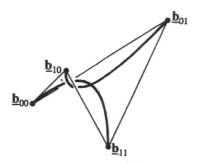

Figure 11.2. Bilinear patches: the two diagonal curves.

is quadratic in t. Note that the two straight lines through a point are a special case of this result.

We thus have: straight lines in the domain of a bilinear patch are mapped to conics. Since these conics are planar curves, they may be viewed as the intersection curves between the patch and a set of planes. Now a natural question is: are *all* planar sections of a bilinear patch conic sections? They are indeed, although we give no proof here. Note that *every* plane intersects the bilinear patch, since it intersects every plane in the two families of straight lines!

Thus a rational bilinear patch is a *quadric*, since a quadric is defined by the fact that any of its planar sections yields a conic. According to section 13.1, it is either a hyperbolic paraboloid — then it is tangent to the plane at infinity; or it is a hyperboloid of one sheet — then it intersects the plane at infinity. The special case of all $w_{ij} = 1$, i.e., the integral case, yields the hyperbolic paraboloid.

A look at the "diagonals" clarifies this observation. Let $\underline{\mathbf{d}}_1(t) = \underline{\mathbf{x}}(t, t)$ and $\underline{\mathbf{d}}_2(t) = \underline{\mathbf{x}}(t, 1 - t)$ be the two diagonal curves. Each of these curves is a conic, and has the following Bézier points:

$$\underline{\mathbf{d}}_1: \qquad \underline{\mathbf{b}}_{00}, (\underline{\mathbf{b}}_{01} + \underline{\mathbf{b}}_{10})/2, \underline{\mathbf{b}}_{11}$$
$$\underline{\mathbf{d}}_2: \qquad \underline{\mathbf{b}}_{01}, (\underline{\mathbf{b}}_{00} + \underline{\mathbf{b}}_{11})/2, \underline{\mathbf{b}}_{10}.$$

It is easy to verify that the two diagonal curves pass through the point $\underline{\mathbf{x}}(\frac{1}{2}, \frac{1}{2})$ — see Fig. 11.2 for an illustration.

But they meet one more time: we also have that

$$\underline{\mathbf{d}}_1(\infty) \hat{=} \underline{\mathbf{d}}_2(\infty) \hat{=} \underline{\mathbf{b}}_{00} - \underline{\mathbf{b}}_{01} - \underline{\mathbf{b}}_{10} + \underline{\mathbf{b}}_{11},$$

which is verified by a simple exercise in algebra.[2]

---

[2]It is interesting to note that this equals the mixed partial derivative $\underline{\mathbf{x}}_{uv}$, or the twist of the bilinear patch! See Section 11.3

After projecting into affine space, we have a rational patch (11.1) with weights w_{ij}, and the diagonal curves will be of one of the three conic types. They can either be both parabolas or they can be one hyperbola and one ellipse. In the first case, the patch *touches* the plane at infinity, and we have the case of a hyperbolic paraboloid. In the second case, the plane at infinity is *intersected*, namely by the diagonal hyperbola, and we have a hyperboloid of one sheet.

In analogy to Section 7.14, one may also distribute points on a bilinear patch and interpret them as the control net of a Bézier patch. With appropriate selection of weights, the patch will actually be linearly parametrized — see [31].

11.2   Bézier patches

A rational rectangular Bézier patch is given by

$$\mathbf{x}(u, v) = \frac{\sum_{i=0}^{m} \sum_{j=0}^{n} w_{ij} \mathbf{b}_{ij} B_i^m(u) B_j^n(v)}{\sum_{i=0}^{m} \sum_{j=0}^{n} w_{ij} B_i^m(u) B_j^n(v)}; \quad 0 \le u, v \le 1. \tag{11.3}$$

It is the image of an integral projective patch:

$$\underline{\mathbf{x}}(u, v) = \sum_{i=0}^{m} \sum_{j=0}^{n} \underline{\mathbf{b}}_{ij} B_i^m(u) B_j^n(v). \tag{11.4}$$

The points $\underline{\mathbf{b}}_{ij}$ or $\mathbf{b}_{ij}$ are called the *control points* of the surface; they form the *control net*. The w_{ij} of the rational patch are called *weights*. The control points are arranged in a rectangular pattern; here is the bicubic case:

$$\underline{\mathbf{b}}_{00}\underline{\mathbf{b}}_{10}\underline{\mathbf{b}}_{20}\underline{\mathbf{b}}_{30}$$
$$\underline{\mathbf{b}}_{01}\underline{\mathbf{b}}_{11}\underline{\mathbf{b}}_{21}\underline{\mathbf{b}}_{31}$$
$$\underline{\mathbf{b}}_{02}\underline{\mathbf{b}}_{12}\underline{\mathbf{b}}_{22}\underline{\mathbf{b}}_{32}$$
$$\underline{\mathbf{b}}_{03}\underline{\mathbf{b}}_{13}\underline{\mathbf{b}}_{23}\underline{\mathbf{b}}_{33}$$

Bézier patches are well understood, and there is a vast amount of literature on them, see [57], [90]. Instead of rederiving every known result for the rational case, we just state some of the fundamental properties:

- If all weights are positive, the rational patch lies in the convex hull of its control net. If we allow the parameters to take on values outside [0, 1], the corresponding surface points will no longer be in this convex hull.

- The four boundary curves of the patch are defined by the control net boundary polygons and their weights.

Figure 11.3. Rational bicubics: left, an integral patch; right: with one weight increased to 10.

- In particular, the four control net corners lie on the patch.

- Every isoparametric curve $u = const$ or $v = const$ is a rational curve of degree m or n, respectively.

- The "diagonal" $\mathbf{x}(t, t)$ is a rational curve of degree $m + n$.

- A straight line may have up to $2mn$ intersections with the patch. This number is referred to as its *algebraic degree*.

- A rational patch is *degree elevated* by degree elevating all its rows or columns according to Section 7.3.

- A rational patch is *subdivided* by interpreting all of its control polygon rows (or columns) as rational curve control polygons and then subdividing them, following the subdivision process outlined in Section 7.10.

Changing the weights of a rational patch has the familiar effect of pulling the patch toward the corresponding control (weight increase) or pushing it away from it (weight decrease). Fig. 11.3 shows this effect.

A more subtle look at weight changes is provided by Fig. 11.4. In the figure, we use *isophotes* to highlight patch shape changes. Isophotes are areas on a surface where the normal vector does not differ by more than a prescribed tolerance from a fixed vector. Isophotes[3] are a very effective tool for detecting shape features in a surface — much more so than shaded images. They take some time to get used to, but once mastered, they are an indispensable tool for surface design.

---

[3]Isophotes are a special case of *reflection lines*, which simulate the reflection of a straight line in a surface.

Figure 11.4. Rational bicubics: top left: an integral patch. Top right: a set of isophotes. Bottom left: isophotes after $w_1 1$ was changed to 2.0. Bottom right: the difference plot of the two sets of isophotes.

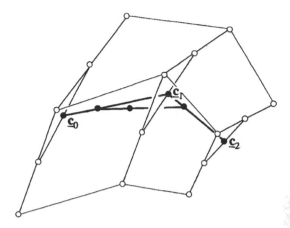

Figure 11.5. Surface evaluation: in this example, three de Casteljau algorithms are performed in the $v-$ direction, then one in the $u-$ direction.

11.3 De Casteljau algorithms and blossoms

The de Casteljau algorithm for Bézier curves carries over to surfaces in two different ways. For a description, it is more convenient to stay in projective space.

First, let us rewrite (11.4) as

$$\underline{\mathbf{x}}(u,v) = \sum_{i=0}^{m} \Big[\sum_{j=0}^{n} \underline{\mathbf{b}}_{ij} B_j^n(v) \Big] B_i^m(u).$$

Denoting the terms in square brackets by $\underline{\mathbf{c}}_i(v)$, we can now formulate our surface evaluation algorithm as:

1. For each i, only consider the ith column of control points and treat them as a curve polygon. Perform a curve de Casteljau algorithm with respect to v, yielding a point $\underline{\mathbf{c}}_i(v)$.

2. Interpret the resulting points $\underline{\mathbf{c}}_i(v)$ as a control polygon for a Bézier curve, and perform one more de Casteljau algorithm with respect to u. The result is the desired surface point.

Fig. 11.5 illustrates this approach.

The points $\underline{\mathbf{c}}_i(v); i = 0, \ldots, m$ are the control points of the *isoparametric curve* corresponding to the fixed parameter value v.

We could have organized our approach differently: we might have started by evaluating in the $u-$ direction first, and then in the $v-$direction. That way, we could have found a tangent in the $v-$direction of the surface. Many applications of surface evaluations do not only need the point on the surface, but also its tangent plane. We then need tangents in both the $u-$ and the $v-$ direction. Instead of running our algorithm twice, we reorganize our evaluation strategy by employing *blossoms*.

The blossom

$$\underline{\mathbf{b}}[u_1, \ldots, u_m \mid v_1, \ldots, v_n] \tag{11.5}$$

of a rectangular patch is defined by first computing the blossom value $[u_1, \ldots, u_m]$ for every row, and then using these values as Bézier control points for a blossom value $[v_1, \ldots, v_n]$ in the $v-$direction. Of course, we could have computed it the other way around: first v, then u, as well!

In case all u_i equal a value u and all v_j equal a value v, then the blossom yields a point on the surface:

$$\underline{\mathbf{x}}(u, v) = \underline{\mathbf{b}}[u^{<m>} \mid v^{<n>}].$$

The Bézier points may also be obtained as special blossom values:

$$\underline{\mathbf{b}}_{i,j} = \underline{\mathbf{b}}[0^{<i>}, 1^{<m-i>} \mid 0^{<j>}, 1^{<n-j>}], \tag{11.6}$$

almost exactly as in the curve case!

An *isoparametric curve* in the $u-$direction may also be formulated in terms of blossoms:

$$\underline{\mathbf{c}}_i = \underline{\mathbf{b}}[0^{<m-i>}, 1^{<i>} \mid v^{<n>}]. \tag{11.7}$$

The corresponding rational curve is obtained in the usual way: with w_i being the fourth coordinate of $\underline{\mathbf{c}}_i$, the rational curve's control points $\mathbf{c}_i$ are given by $\mathbf{c}_i = \underline{\mathbf{c}}_i/w_i$, and their weights are the w_i.

Needless to say, we can compute the control points of isoparametric curves in the $v-$direction in a completely analogous way!

11.4   The weight points

We may form the averages $\mathbf{p}_{i,j}$ in the $i-$ direction of any two adjacent control points:

$$\mathbf{p}_{i,j} = \frac{w_{i,j}\mathbf{b}_{i,j} + w_{i+1,j}\mathbf{b}_{i+1,j}}{w_{i,j} + w_{i+1,j}}$$

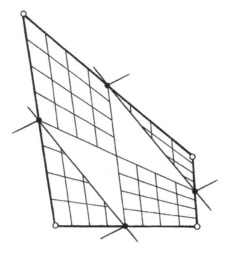

Figure 11.6. Weight points: neighboring control points (open) give rise to coplanar weight points (solid).

and similarly the averages $q_{i,j}$ in the j-direction:

$$q_{i,j} = \frac{w_{i,j}b_{i,j} + w_{i,j+1}b_{i,j+1}}{w_{i,j} + w_{i,j+1}}.$$

These *weight points* are shown in Fig. 11.6, there for the special case $m = n = 1$ and all $w_{i,j} = 1$. This characterizes a hyperbolic paraboloid — note that for this case, the weight points actually form a parallelogram.

Any subquadrilateral has four weight points on its edges. Although the control weights and control points are arbitrary, the weight points always satisfy one condition: they are *coplanar*.[4] This may be surprising in an affine context, but it is close to trivial in a projective one: there we have

$$\underline{p}_{i,j} = \underline{b}_{i,j} + \underline{b}_{i+1,j}, \quad \underline{q}_{i,j} = \underline{b}_{i,j} + \underline{b}_{i,j+1}.$$

Assigning coordinates[5]

$$\underline{b}_{i,j} = \begin{bmatrix} 1 \\ 0 \\ 0 \\ 0 \end{bmatrix}, \underline{b}_{i+1,j} = \begin{bmatrix} 0 \\ 1 \\ 0 \\ 0 \end{bmatrix}, \underline{b}_{i,j+1} = \begin{bmatrix} 0 \\ 0 \\ 1 \\ 0 \end{bmatrix}, \underline{b}_{i+1,j+1} = \begin{bmatrix} 0 \\ 0 \\ 0 \\ 1 \end{bmatrix},$$

---

[4]This was first pointed out to me by H. Pottmann.

[5]Recall that we may always do this without loss of generality, as long as the four points are not coplanar — in which case we have nothing to prove.

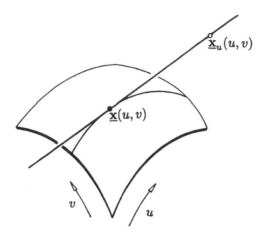

Figure 11.7. Partials: a $u-$ partial of a surface.

all we need to do is check that the determinant $|\underline{\mathbf{P}}_{i,j}, \underline{\mathbf{P}}_{i,j+1}, \underline{\mathbf{q}}_{i,j}, \underline{\mathbf{q}}_{i+1,j}|$
vanishes. This is a straightforward exercise in algebra.

We have thus lost a useful property of rational Bézier curves: there, we could assign weight points in an intuitive and geometric way — now, the weight points have to satisfy constraints and can thus not be used as independent input.

11.5   Derivatives and continuity

We may use isoparametric curves to define certain derivatives of a surface. Let $\underline{c}_0, \ldots, \underline{c}_m$ be the control points of the isoparametric curve corresponding to a fixed valued of v. We may evaluate this curve using the de Casteljau algorithm. The final step (in the $u-$direction) yields the derivative of the isoparametric curve, and thus a derivative of the surface. It is called the $u-$partial $\underline{\mathbf{x}}_u(u, v)$ of the surface:

$$\underline{\mathbf{x}}_u(u, v) = \frac{\partial \underline{\mathbf{x}}(u, v)}{\partial u} = \frac{\mathrm{d}}{\mathrm{d}u} \sum_{i=0}^{m} \underline{c}_i(v) B_i^m(u). \qquad (11.8)$$

Keep in mind that in $I\!\!P^3$, this derivative corresponds to a point! See Fig. 11.7.

Figure 11.8. The bilinear osculant: it agrees with the given surface in $u-$ and $v-$partials.

Expanding (11.8), we obtain

$$\underline{\mathbf{x}}_u(u,v) = m \sum_{i=0}^{m-1} [\underline{\mathbf{c}}_{i+1} - \underline{\mathbf{c}}_i] B_i^{m-1}(u). \tag{11.9}$$

In a completely analogous way, we could have computed the $v-$ partial $\underline{\mathbf{x}}_v$.

Another way to compute first partials is as follows. The *bilinear osculant* at $\underline{\mathbf{x}}(u,v)$ is the bilinear patch given by four Bézier points $\underline{\mathbf{a}}_{i,j}$, which we express in blossom notation:

$$\begin{bmatrix} \underline{\mathbf{a}}_{0,0} & \underline{\mathbf{a}}_{0,1} \\ \underline{\mathbf{a}}_{1,0} & \underline{\mathbf{a}}_{1,1} \end{bmatrix} = \begin{bmatrix} \underline{\mathbf{b}}[u^{<m-1>},0 \mid v^{<n-1>},0] & \underline{\mathbf{b}}[u^{<m-1>},0 \mid v^{<n-1>},1] \\ \underline{\mathbf{b}}[u^{<m-1>},1 \mid v^{<n-1>},0] & \underline{\mathbf{b}}[u^{<m-1>},1 \mid v^{<n-1>},1] \end{bmatrix},$$

$$\tag{11.10}$$

see Fig. 11.8. Of course, it does not matter in which order the blossoms are computed.

Fig. 11.9 illustrates one strategy to compute the bilinear osculant for the case $m = 3$, $n = 2$.

Figure 11.9. The bilinear osculant: several de Casteljau steps for its computation are shown.

The bilinear osculant with these four control points agrees with the original surface in u−partial, v−partial, and twist. In particular, it therefore agrees with the surface in its tangent plane.

Higher derivatives in projective space[6] may also be computed using the blossoming technique:

$$\frac{\partial^{r+s}}{\partial u^r \partial v^s} \mathbf{x}(u,v) \hat{=} \underline{\mathbf{b}}[u^{<m-r>}, \infty^{<r>} \mid v^{<n-s>}, \infty^{<s>}]. \qquad (11.11)$$

Compare with (7.38)!

A point on a *rational* patch may be computed by performing the above projective evaluation algorithms and then projecting the result into affine space, or by projecting every step of those algorithms into affine space. In particular, we may compute the rational bilinear osculant to the patch: it is given by the four points

$$\mathbf{b}_{0,0}^{m-1,n-1}, \mathbf{b}_{0,1}^{m-1,n-1}, \mathbf{b}_{1,0}^{m-1,n-1}, \mathbf{b}_{1,1}^{m-1,n-1}$$

and their respective weights

$$w_{0,0}^{m-1,n-1}, w_{0,1}^{m-1,n-1}, w_{1,0}^{m-1,n-1}, w_{1,1}^{m-1,n-1}.$$

We may now use this rational bilinear osculant to obtain partials of the rational patch:

$$\mathbf{x}_u(u,v) = m \frac{w_{0,0}^{m-1,n} w_{1,0}^{m-1,n}}{[w_{0,0}^{m,n}]^2} \left[\mathbf{b}_{1,0}^{m-1,n} - \mathbf{b}_{0,0}^{m-1,n} \right] \qquad (11.12)$$

and

$$\mathbf{x}_v(u,v) = n \frac{w_{0,0}^{m,n-1} w_{0,1}^{m,n-1}}{[w_{0,0}^{m,n}]^2} \left[\mathbf{b}_{0,1}^{m,n-1} - \mathbf{b}_{0,0}^{m,n-1} \right] \qquad (11.13)$$

Fig. 11.10 explains the notation.

If we want the tangent plane (or, equivalently, the normal) of the corresponding rational surface in affine space, we simply take the cross product of $\mathbf{x}_u$ and $\mathbf{x}_v$. For a general formula of partials of a rational Bézier patch, see [154] or [136].

A Bézier patch $\mathbb{P}^3$ is the map of a quadrilateral into $\mathbb{P}^3$. The quadrilateral, the domain of the patch, defines a Moebius net, see Section 1.5. If we consider *two* Bézier patches which share a common boundary curve, we may ask for conditions that make the resulting composite surface C^1. For this to be meaningful, the domain quadrilaterals of each patch must define the

---

[6]Recall: these are *points*!

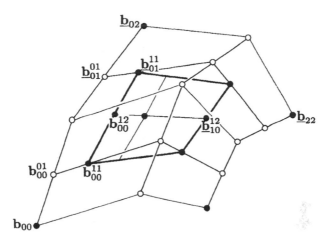

Figure 11.10. Partials: the involved control points on the bilinear osculant.

same Moebius net of $I\!P^2$ — this is in analogy to the requirement of equivalent reference frames when we discussed C^1 continuity of curves in Section 6.1. Suppose the domains of our two patches $\underline{x}$ and $\underline{y}$ are aligned as in Fig. 11.11, also assume that $\underline{x}$ is defined by a control net $\underline{\mathbf{b}}_{i,j}; i = 0, \ldots, m; j = 0, \ldots, n$ and that $\underline{y}$ is defined by $\underline{\mathbf{b}}_{i,j}; i = m, \ldots, 2m; j = 0, \ldots, n$. Then the condition of equal u–partials for every v along the common boundary $u = u_1$ yields

$$\frac{d}{du}\underline{x}(u_1, v) = \frac{d}{du}\underline{y}(u_1, v). \tag{11.14}$$

Inserting this into (11.9), and after comparing coefficients of $B_j^n(v)$, we obtain as a C^1 condition: all rows of control points must define composite Bézier curves which are C^1 with respect to the intervals u_0, u_1, u_2. Thus (referring to (6.1)), we obtain

$$\underline{\mathbf{b}}_{m,j} = \frac{\Delta_1}{\Delta_0 + \Delta_1}\underline{\mathbf{b}}_{m-1,j} + \frac{\Delta_0}{\Delta_0 + \Delta_1}\underline{\mathbf{b}}_{m+1,j}; j = 0, \ldots, n \tag{11.15}$$

with $\Delta_0 = u_1 - u_0$ and $\Delta_1 = u_2 - u_1$.

Note that C^1 continuity in $I\!P^3$ implies C^1 continuity for the corresponding rational patches in affine space. For more details on this, see [56].

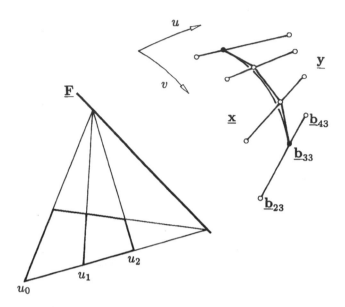

Figure 11.11. C^1 continuity: two patch domains are part of *one* Moebius net of $I\!P^2$. The relevant parts of the control nets (range) are also shown.

11.6   Reparametrization

In the curve case, we could change the parametrization by a Moebius transformation without changing the curve shape; the same is true now. The parameters u and v are independent variables; we may subject each of them to a Moebius transformation independently. If the reparametrization constants (see Sec. 7.5) are c for the parameter u and d for v, then the new patch equation is

$$\underline{\mathbf{x}}(\hat{u}, \hat{v}) = \sum_{i=0}^{m} \sum_{j=0}^{n} c^i d^j \underline{\mathbf{b}}_{i,j} B_i^m(\hat{u}) B_j^n(\hat{v}). \qquad (11.16)$$

The corresponding rational case is now trivial:

$$\mathbf{x}(\hat{u}, \hat{v}) = \frac{\sum_{i=0}^{m} \sum_{j=0}^{n} c^i d^j w_{i,j} \mathbf{b}_{i,j} B_i^m(\hat{u}) B_j^n(\hat{v})}{\sum_{i=0}^{m} \sum_{j=0}^{n} c^i d^j w_{i,j} B_i^m(\hat{u}) B_j^n(\hat{v})} \qquad (11.17)$$

Fig. 11.12 shows the effect of such a reparametrization on a grid of isoparametric curves of the surface from Fig. 11.4.

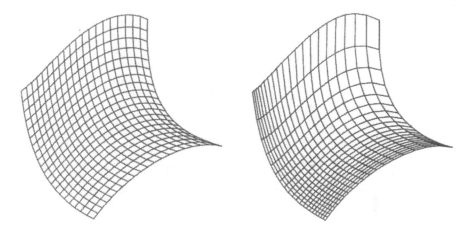

Figure 11.12. Surface reparametrization: left, an integral patch, right: after a reparametrization with $c = 2$, $d = 3$.

11.7 Rectangular NURBS surfaces

Rational Bézier patches were obtained as follows: construct a tensor product surface in projective space, then project into affine space, thus creating a rational surface. The story does not change for B-splines: a projective, piecewise polynomial tensor product B-spline surface is given by

$$\underline{s}(u, v) = \sum_{i=0}^{p} = \sum_{j=0}^{q} \underline{d}_{i,j} N_i^m(u) N_j^n(v), \qquad (11.18)$$

where the $N_i^m(u)$ and $N_j^n(v)$ are the univariate B-spline basis functions from Chapter 9. The knot sequence in the $u-$direction is defined by $\{u_0, \ldots, u_K\}$; in the $v-$direction, we have $\{v_0, \ldots, v_L\}$.

After projection into affine space, we obtain the NURB surface

$$s(u, v) = \frac{\sum_{i=0}^{p} \sum_{j=0}^{q} v_{i,j} \mathbf{d}_{i,j} N_i^m(u) N_j^n(v)}{\sum_{i=0}^{p} \sum_{j=0}^{q} v_{i,j} N_i^m(u) N_j^n(v)}; \quad \mathbf{d}_{i,j} \in I\!\!E^3. \qquad (11.19)$$

The weights $v_{i,j}$ are the fourth coordinates of the $\underline{d}_{i,j}$ — we do not denote them by $w_{i,j}$ to avoid confusion with the Bézier weights. As usual, the $\{\mathbf{d}_{i,j}\}$ are called *control points*, forming the control net. Fig. 11.13 shows an example.

Figure 11.13. NURB surfaces: top, control net and surface; middle: isophotes; bottom: isophotes, after doubling some weights.

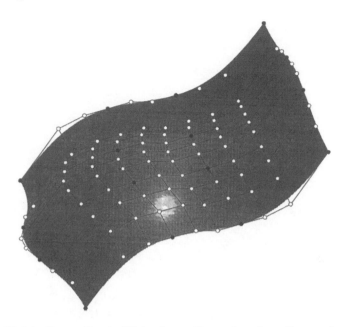

Figure 11.14. Conversion to Bézier form: the surface from the previous figure, but now broken down into rational Bézier patches.

The weights $v_{i,j}$ may be used to fine-tune a given shape. In Fig. 11.13, we use isophotes as above to highlight shape features of a surface. After changing some weights, a new pattern emerges. The two patterns are not markedly different — the only changes are in the right part of the surface. This is a typical situation in surface design: sometimes tiny changes are sufficient to meet a designer's intent.

There are several ways to evaluate NURB surfaces; we discuss the tensor product de Boor algorithm and the conversion to rational Bézier form. In the case of rectangular Bézier patches, we had several options to apply de Casteljau algorithms for the evaluation of a surface point. The same is true now, except that we replace univariate de Casteljau algorithms by univariate de Boor algorithms. One strategy is as follows: first, we determine which intervals u and v are in, i.e., we find I and J such that $u_I \leq u < u_{I+1}$ and $v_J \leq v < v_{J+1}$ Then, we take every row of the projective control net $\{\underline{\mathbf{d}}_{i,j}\}$, interpret it as a B-spline curve control polygon, and use the de Boor algorithm (9.3) to evaluate it at u. Then take all points $\underline{\mathbf{d}}_{I+1,0}^{m,0}, \ldots, \underline{\mathbf{d}}_{I+1,K+n-1}^{m,0}$ found in this manner, interpret them as a control polygon, and evaluate at v. The resulting point $\underline{\mathbf{d}}_{I+1,J+1}^{m,n}(u,v)$ is the desired projective point on the surface, and is mapped into affine space upon division by the fourth coordinate.

Figure 11.15. NURB surfaces: a composite surface.

Conversion to Bézier form follows the same recipe: first convert all projective control net rows into Bézier control polygons, using the blossom approach from Section 9.5. Take each column of the resulting array, interpret it as a B-spline polygon, and convert to Bézier form, again using blossoms. The result, after dividing by the fourth coordinates, is the desired piecewise Bézier net together with its weights. Fig. 11.14 give an example.

Fig. 11.15 shows another composite NURB surface, this time consisting of several surfaces of revolution joined together by toroidal pieces.

11.8 Blossoms

Blossoms for B-spline surfaces in $I\!P^3$ do not offer many surprises if you understand those for Bézier patches. In complete analogy to (11.5), we define

$$\underline{\mathbf{d}}^{\lambda,\Omega}[r_1,\ldots,r_m|s_1,\ldots,s_n], \tag{11.20}$$

indicating that we are using an interval $\lambda = [u_I, u_{I+1}]$ for evaluation in the $u-$ direction and an interval $\Omega = [v_J, v_{J+1}]$ for evaluation in the $v-$ direction. The B-spline control points $\underline{\mathbf{d}}_{I-m+1+k,J-n+1+l}$ which define the patch $u \in \lambda, v \in \Omega$ are then given by

$$\underline{\mathbf{d}}_{I-m+1+k,J-n+1+l} = \mathbf{d}^{\lambda,\Omega}[u_{I-m+1+k},\ldots,u_{I+k},v_{J-n+1+l},\ldots|v_{j+l}],$$

with $k = 0, \ldots, m$ and $l = 0, \ldots, n$. This is not the most effective way to carry out the computation, but certainly the one with the cleanest code!

Conversion to piecewise Bézier form follows the same idea: the Bézier patch corresponding to $u \in \lambda$ and $v \in \Omega$ has control points

$$\mathbf{b}_{i,j}^{\lambda,\Omega} = \mathbf{d}^{\lambda,\Omega}[u_I^{<m-i>}, u_{I+1}^{<i>} | v_J^{<n-j>}, v_{J+1}^{<j>}].$$

In order to compute $u-$ and $v-$ partials, it is easiest to convert to Bézier form and then proceed as in Section 11.5.

11.9   Surfaces of revolution

Surfaces of revolution play an important role in all of engineering and manufacturing. Using integral surface patches, we can only approximate them — rational surfaces allow an exact representation, provided that the *generatrix* is itself a NURB curve. In general, a generatrix is a curve that sweeps out a surface. In our case, we define an arbitrary straight line **L**, call it the *axis* of the surface of revolution, and let the generatrix rotate around the axis, thus sweeping out the surface. Clearly, such surfaces are *Euclidean* objects, since they contain infinitely many circles, the *meridians*.

In most practical applications, the generatrix is planar, and lies in a plane also containing the axis.

Analytically, a surface of revolution is given by

$$\mathbf{x}(u, v) = \begin{bmatrix} r(v) \cos u \\ r(v) \sin u \\ z(v) \end{bmatrix}.$$

For fixed v, an isoparametric line $v = const$ traces out a circle of radius $r(v)$, namely a meridian. Since a circle may be exactly represented by rational quadratic arcs, we may find an exact rational representation of a surface of revolution provided we can represent $r(v), z(v)$ in rational form.

The most convenient way to define a surface of revolution is to prescribe the generatrix, given by

$$\mathbf{g}(v) = [r(v), 0, z(v)]^T$$

and by the axis, in the same plane as **g**. Suppose **g** is given by its control polygon, knot sequence, and weight sequence. We can construct a surface of revolution such that each meridian consists of four rational quadratic arcs, as shown in Figure 5.9. For each vertex of the generating polygon, construct a square (perpendicular to the axis of revolution) as in Figure 5.9.

Figure 11.16. Surfaces of revolution: top, the surface with the generatrix control polygon; middle: one quarter of the surface with its control polygon; bottom: shaded view of whole surface.

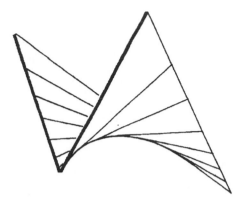

Figure 11.17. Developable surfaces: all tangents to a space curve form a developable surface. The space curve is its edge of regression.

Assign the given weights of the generatrix to the four polygons corresponding to the square's edge midpoints; the remaining weights, corresponding to the squares' corners, are multiplied by $\cos(45°) = \sqrt{2}/2$. In this way, we can represent "classical" surfaces such as cylinders, spheres, or tori. See Chapter 15 for numerical examples, and Figure 11.16 for an illustration.

Note that although the generatrix may be defined over a knot sequence $\{v_j\}$ with only simple knots, this is not possible for the knots of the meridian circles: we have to use double knots, thereby essentially reducing the meridians to the piecewise Bézier form. Even after degree elevation to rational cubic form, we must continue to use double knots, as the curve is still not twice differentiable in projective space.

11.10 Developable surfaces

In Section 7.12, we encountered control plane curves, defined by

$$\underline{\mathbf{B}}(t) = \sum_{i=0}^{n} \underline{\mathbf{B}}_i(t),, \tag{11.21}$$

with $\underline{\mathbf{B}}(t)$ as well as the $\underline{\mathbf{B}}_i$ being planes. As t varies, the planes $\underline{\mathbf{B}}(t)$ envelope a surface, a so-called *developable surface*. (11.21). Each $\underline{\mathbf{B}}(t)$ is tangent to this surface along a whole line; the collection of all these lines is the envelope of the *edge of regression* of the developable surface. The edge of regression is a sharp ridge on the surface; the surface consists of all tangents of the edge of regression. Fig. 11.17 illustrates.

Figure 11.18. Developable surfaces: all tangent planes to a surface along a curve on it form a developable surface.

Another way of producing developable surfaces is as follows. Consider a curve on a parametric surface. At every point of the curve, there is a tangent plane to the surface. All such tangent planes are envelopes of a surface of revolution, as shown in Fig. 11.18.

The reason for the name "developable surface" is a physical analogue: such surfaces may be flattened out, or unwrapped, into the plane without any distortions. More algebraically, their Gaussian curvature vanishes everywhere.[7] For these differential geometry concepts, see [26], [57] or [46].

A warning concerning a common misconception: although developable surfaces are ruled, not all ruled surfaces are developable! An affine example is supplied by the bilinear patch: it is ruled (even doubly ruled), but its Gaussian curvature is negative everywhere, and hence it is not developable.

Let $\underline{\mathbf{b}}(t)$ be a cubic that is the edge of regression of a developable surface $\underline{\mathbf{d}}$. At a given parameter value t, all points of the corresponding tangent are given by the blossoms $\underline{\mathbf{b}}[t, t, s]$, i.e., as s traces out all reals, $\underline{\mathbf{b}}[t, t, s]$ traces out all of the tangent to $\underline{\mathbf{b}}(t)$. Thus every point on $\underline{\mathbf{d}}$ can be written as

$$\underline{\mathbf{d}}(s, t) = \underline{\mathbf{b}}[t, t, s]. \qquad (11.22)$$

If we think of s as being fixed and t varying, then $\underline{\mathbf{b}}[t, t, s]$ defines a conic; comparing with Section 7.10, we see that it is the cubic's *osculant* at $\underline{\mathbf{b}}(s)$. Thus $\underline{\mathbf{d}}$ does not only contain a family of lines, but a family of conics as well! Each of these conics may be thought of as the intersection of $\underline{\mathbf{d}}$ and one of its *osculating planes*, see Section 7.10.

---

[7]The Gaussian curvature of a surface is a measure for the "roundness" of a surface at a given point. Every point on a hyperboloid has negative Gaussian curvature; every point on an ellipsoid has positive Gaussian curvature, and every point on a cylinder has zero Gaussian curvature.

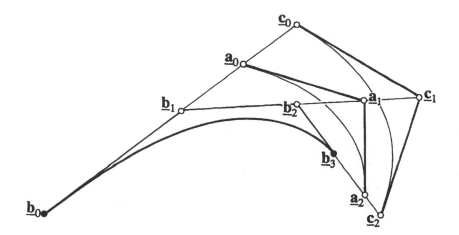

Figure 11.19. Conics on developables: two conic segments together with the edge of regression.

Let us now consider two of these conics, $\underline{a}(t)$ and $\underline{c}(t)$, say. Each of them, illustrated in Fig. 11.19, is given by a set of control points $\{\underline{a}_i\}$ and $\{\underline{c}\}_i$, respectively, plus a shoulder tangent each (or, alternatively, two weight points each).

Let $\underline{a}(t)$ and $\underline{c}(t)$ be the cubic's osculants at $t = \alpha$ and $t = \gamma$, respectively. We can write their control points as blossom values:

$$\underline{a}_0 = \underline{b}[0,0,\alpha], \underline{a}_1 = \underline{b}[0,1,\alpha], \underline{a}_2 = \underline{b}[1,1,\alpha]$$

and

$$\underline{c}_0 = \underline{b}[0,0,\gamma], \underline{c}_1 = \underline{b}[0,1,\gamma], \underline{c}_2 = \underline{b}[1,1,\gamma].$$

Thus the cross ratios formed by four points on the edges of $\underline{b}$'s control polygon are all equal:

$$\mathrm{cr}(\underline{b}_0, \underline{b}_1, \underline{a}_0, \underline{c}_0) = \mathrm{cr}(\underline{b}_1, \underline{b}_2, \underline{a}_1, \underline{c}_1) = \mathrm{cr}(\underline{b}_2, \underline{b}_3, \underline{a}_2, \underline{c}_2). \qquad (11.23)$$

After this investigation of properties of developable surfaces, we now turn to a more practical question: given two arbitrary conic segments $\underline{a}(t)$ and $\underline{c}(t)$, can we find a cubic developable surface $\underline{d}(s,t)$ of the form

$$\underline{d}(s,t) = (1-s)\underline{a}(t) + s\underline{c}(t)? \qquad (11.24)$$

In general, the answer is negative, as Fig. 11.19 illustrates: we would have to find $\underline{d}$'s edge of regression $\underline{b}(t)$. Then the four points $\underline{a}_0, \underline{c}_0, \underline{a}_1, \underline{c}_1$

have to be coplanar, namely they have to be in $\underline{\mathbf{b}}$'s osculating plane at $\underline{\mathbf{b}}_0$, spanned by $\underline{\mathbf{b}}_0, \underline{\mathbf{b}}_1, \underline{\mathbf{b}}_2$. Similarly, the four points $\underline{\mathbf{a}}_1, \underline{\mathbf{c}}_1, \underline{\mathbf{a}}2, \underline{\mathbf{c}}_2$ must lie in $\underline{\mathbf{b}}$'s osculating plane at $\underline{\mathbf{b}}_3$.

Suppose then that our two conics meet these coplanarity conditions. Our first aim is to find the control polygon for the edge of regression. We immediately find $\underline{\mathbf{b}}_1$ as the intersection of the lines through $\underline{\mathbf{a}}_0, \underline{\mathbf{c}}_0$ and $\underline{\mathbf{a}}_1, \underline{\mathbf{c}}_1$ and $\underline{\mathbf{b}}_2$ as the intersection of the two lines through $\underline{\mathbf{a}}_1, \underline{\mathbf{c}}_1$ and $\underline{\mathbf{a}}_2, \underline{\mathbf{c}}_2$.

Next, we compute the cross ratio formed by the four points $\underline{\mathbf{b}}_1, \underline{\mathbf{b}}_2, \underline{\mathbf{a}}_1, \underline{\mathbf{c}}_1$. We can then find $\underline{\mathbf{b}}_0$ and $\underline{\mathbf{b}}_3$ from the cross ratio condition (11.23), thus having found the desired control polygon.

We have not yet determined the weight points of the two conics. For this task, recall that two conic osculants to a cubic must share a common tangent, as shown in Fig. 7.15. This tangent must then be the intersection of the two planes that contain the conics, a condition that will not be met by two arbitrary conics. If $\underline{\mathbf{q}}_i^a$ and $\underline{\mathbf{q}}_i^c$ are the weight points of the two conics, and $\underline{\mathbf{q}}_i$ are those of the cubic, it is not hard to show that $\underline{\mathbf{q}}_0, \underline{\mathbf{q}}_1, \underline{\mathbf{q}}_0^a, \underline{\mathbf{q}}_0^c$ are collinear. Similarly, $\underline{\mathbf{q}}_1, \underline{\mathbf{q}}_2, \underline{\mathbf{q}}_1^a, \underline{\mathbf{q}}_1^c$ are collinear. Thus we can prescribe $\underline{\mathbf{q}}_0^a$ and $\underline{\mathbf{q}}_0^c$ and determine the remaining $\underline{\mathbf{q}}_1^a, \underline{\mathbf{q}}_1^c$.

This construction relies heavily upon properties that the given conics must satisfy. In real applications, this will rarely be the case, and approximative solutions must be sought, as outlined in [125].

Leopoldsleder and Pottmann [101] investigated the construction of developable surfaces as *cone splines*. If we define two space curves as C^1 conics, with the same number of conics in each curve, then the ruled surface between them is a piecewise conic surfaces, and hence developable. The problem thus boils down to finding appropriate conic spline approximations to boundary curves which might not have been given in this form initially.

Developable surfaces play an important role in several applications. In the stamping process of sheet metals, so-called *binder surfaces* are used to clamp the sheet metal around the die. Initially, the sheet falls onto the binder surface without touching the die. In order to touch the binder surface everywhere without wrinkling, the binder must be developable.[8]

Another application is in the shipbuilding industry, where ship hulls are assembled from rolled sheets — again forming developable surfaces.

More literature on this topic: Aumann [8] Ravani and Boddulori [21], Frey [19], Pottmann and Farin [125], Redont [129], Kim [94].

---

[8]In practice, it suffices that the binder is close to being developable.

11.11   Problems

1. Show that the diagonal curves of a rational bilinear patch cannot both be elliptic or both hyperbolic.

2. Verify the following:
   a) a hyperboloid of one sheet is obtained as a surface of revolution with the hyperbola as the generatrix and the axis of symmetry between the two branches as axis.
   b) For a hyperboloid of two sheets, the generatrix is again the hyperbola, but the axis is the common axis of symmetry of both branches.

3. Show that the planar weight point quadrilaterals from Section 11.4 are parallelograms if all weights are equal. Note that this is strictly an affine phenomenon.

4. Show that every planar section of a bilinear patch in $I\!P^3$ is a conic section.

12

Rational Bézier Triangles

Bézier triangles, short for triangular Bézier patches, actually owe their discovery to P. de Casteljau from Citröen, who outlined their theory in the two technical reports [42] and [43]. Here, we are interested in their rational counterparts, but for the sake of simplicity, we start with an outline of integral, polynomial Bézier triangles in projective space. These will then be projected onto their rational counterparts.

Triangular patches enjoy some popularity among CAGD researchers, although they have not been included in standards such as IGES. They are useful when it comes to modeling complex shapes that do not lend themselves to a rectangular topology. *Rational* Bézier triangles have not been investigated very thoroughly yet; some pointers to the literature include [25], [84].

12.1 Projective Bézier triangles

Before exploring rational Bézier triangles, we introduce their projective pre-images — triangular polynomial patches in $I\!\!P^3$, expressed in Bernstein-Bézier form. They behave much like their integral (polynomial) counterparts. For more complete information about integral Bézier triangles, the reader is referred to [51] or [57].

A Bézier patch in $I\!\!P^3$ is defined by a control net similar to those of tensor product patches, except that it comes in a triangular arrangement. It is convenient to give each control point three subscripts, and a typical control net then looks like this:

$$\begin{array}{c}
\underline{b}_{040} \\
\underline{b}_{031}\,\underline{b}_{130} \\
\underline{b}_{022}\,\underline{b}_{121}\,\underline{b}_{220} \\
\underline{b}_{013}\,\underline{b}_{112}\,\underline{b}_{211}\,\underline{b}_{310} \\
\underline{b}_{004}\,\underline{b}_{103}\,\underline{b}_{202}\,\underline{b}_{301}\,\underline{b}_{400}
\end{array}$$

Note how all subscripts sum to four — this indicates that we are dealing with the control net of a degree four, or quartic, patch. For other degrees n, the arrangement is similar, and the control point subscripts always sum to n. The control points $\underline{b}_i = \underline{b}_{i,j,k}$ are points in $I\!\!P^3$, thus having four coordinates x, y, z, w each.

Consider an arbitrary (nondegenerate) triangle in the projective plane, which, together with a fourth point, defines a projective reference frame and thus projective coordinates $\underline{u} = [u, v, w]^T$.[1] We give the triangle vertices the coordinates $e1 = [1, 0, 0]^T, e2 = [0, 1, 0]^T, e3 = [0, 0, 1]^T$, and the fourth point the coordinates $[1/3, 1/3, 1/3]^T$. A triangular patch $\underline{b}(\underline{u})$ is a polynomial map of the projective plane into $I\!\!P^3$, such that a point $\underline{x}$ on the triangular patch is expressed as

$$\underline{x} = \underline{b}^n(\underline{u}) = \underline{b}_0^n(\underline{u}) = \sum_{|j|=n} \underline{b}_j B_j^n(\underline{u}), \qquad (12.1)$$

where the $B_i^n(\underline{u})$ are trivariate[2] Bernstein polynomials:

$$B_i^n(\underline{u}) = B_{i,j,k}^n(u, v, w) = \frac{n!}{i!j!k!} u^i v^j w^k; \quad |i| = n. \qquad (12.2)$$

Figure 12.1 gives an example of a cubic patch with its control net.

Bernstein polynomials satisfy the following recursion:

$$B_i^n(\underline{u}) = u B_{i-e1}^{n-1}(\underline{u}) + v B_{i-e2}^{n-1}(\underline{u}) + w B_{i-e3}^{n-1}(\underline{u}); \quad |i| = n, \qquad (12.3)$$

---

[1]The third projective coordinate, w, for $I\!\!P^2$ is named the same as the fourth coordinate for $I\!\!P^3$. As long as there is no danger of confusion, we will use this slightly ambiguous notation.

[2]Note that we are still dealing with *surfaces*, since the u, v, w are coordinates of a point in the plane!

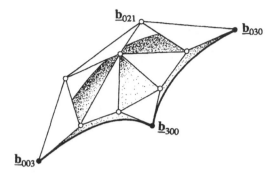

Figure 12.1. Bézier triangles: a cubic patch with its control net.

which can be used to define the de Casteljau algorithm for the evaluation of Bezier triangles:

de Casteljau algorithm:

Given: a triangular array of points $\underline{\mathbf{b}}_i \in I\!\!P^3; |\mathbf{i}| = n$ and a point in $I\!\!P^3$ with coordinates $\underline{\mathbf{u}}$.

Set:

$$\underline{\mathbf{b}}_i^r(\underline{\mathbf{u}}) = u\underline{\mathbf{b}}_{i+e1}^{r-1}(\underline{\mathbf{u}}) + v\underline{\mathbf{b}}_{i+e2}^{r-1}(\underline{\mathbf{u}}) + w\underline{\mathbf{b}}_{i+e3}^{r-1}(\underline{\mathbf{u}}) \tag{12.4}$$

where

$$r = 1, \ldots, n \text{ and } |\mathbf{i}| = n - r$$

and $\underline{\mathbf{b}}_i^0(\underline{\mathbf{u}}) = \underline{\mathbf{b}}_i$. Then $\underline{\mathbf{b}}_0^n(\underline{\mathbf{u}})$ is the point with parameter value $\underline{\mathbf{u}}$ on the *Bézier triangle* $\underline{\mathbf{b}}^n$.

Fig. 12.2 shows the quadratic case.

This algorithm is a direct generalization of the univariate de Casteljau algorithm, and shares many properties with it. In particular, the three intermediate points $\underline{\mathbf{b}}_{e1}^{n-1}, \underline{\mathbf{b}}_{e2}^{n-1}, \underline{\mathbf{b}}_{e3}^{n-1}$, span the tangent plane of the patch at $\underline{\mathbf{b}}(\underline{\mathbf{u}})$.

The boundary curves of a triangular patch are determined by the boundary control vertices (having at least one zero as a subscript). For example, a point on the boundary curve $\underline{\mathbf{b}}^n(u, 0, w)$ is generated by

$$\underline{\mathbf{b}}_i^r(u, 0, w) = u\underline{\mathbf{b}}_{i+e1}^{r-1} + w\underline{\mathbf{b}}_{i+e3}^{r-1},$$

which is the univariate de Casteljau algorithm for Bézier curves.

In analogy to the weight points for Bézier curves, we may define

$$\underline{\mathbf{q}}_i = \underline{\mathbf{b}}_{i+e1} + \underline{\mathbf{b}}_{i+e2} + \underline{\mathbf{b}}_{i+e3}; \ |\mathbf{i}| = n - 1.$$

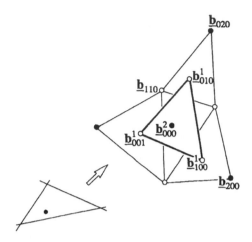

Figure 12.2. The de Casteljau algorithm: the quadratic case. The tangent plane at the computed point is highlighted.

These points are only defined in the "upright" subtriangles, see Fig. 12.3 for an illustration. In the curve case, we could give each of the weight points an arbitrary position on its control polyon leg; this is not possible any more: changing just one of the weight points would result in conflicts with neighboring "upside down" subtriangles.

G. Albrecht has resolved this problem: if the weight points are assigned in a spiral pattern, no conflicts arise. See Fig. 12.4 and [6].

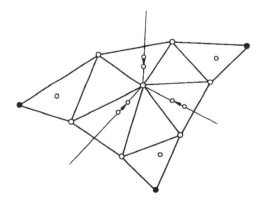

Figure 12.3. Weight points: these may be defined in all "upright" subtriangles. If the increase the center weight, three weight points will move towards it.

Figure 12.4. Weight points: a spiral pattern yields to a consistent patch definition. Solid: Bézier points; open: weight points.

12.2   Degree elevation

It is possible to write $\underline{\mathbf{b}}^n$ as a Bézier triangle of degree $n + 1$:

$$\sum_{|\mathbf{i}|=n} \underline{\mathbf{b}}_{\mathbf{i}} B_{\mathbf{i}}^n(\mathbf{u}) = \sum_{|\mathbf{i}|=n+i} \underline{\mathbf{b}}_{\mathbf{i}}^{(1)} B_{\mathbf{i}}^{n+1}(\mathbf{u}). \tag{12.5}$$

The control points $\underline{\mathbf{b}}_{\mathbf{i}}^{(1)}$ are obtained from

$$\underline{\mathbf{b}}_{\mathbf{i}}^{(1)} = \frac{1}{n+1}[i\underline{\mathbf{b}}_{\mathbf{i}-\mathbf{e1}} + j\underline{\mathbf{b}}_{\mathbf{i}-\mathbf{e2}} + k\underline{\mathbf{b}}_{\mathbf{i}-\mathbf{e3}}]. \tag{12.6}$$

For a proof, we multiply the left hand side of (12.5) by $u + v + w$ and compare coefficients of like powers. Figure 12.5 illustrates the case $n = 2$.

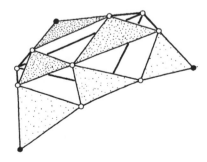

Figure 12.5. Degree elevation: a quadratic control net together with the equivalent cubic net (shaded).

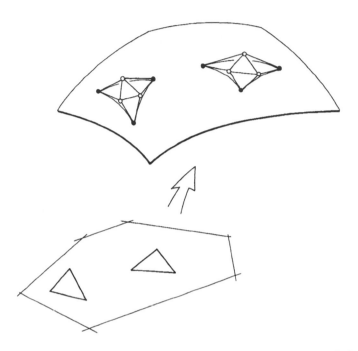

Figure 12.6. Blossoms: two different representations of a globally defined surface.

12.3 Blossoms and derivatives

The *blossom* of a triangular patch is obtained by feeding n possibly different arguments into the de Casteljau algorithm; it is denoted by $\underline{\mathbf{b}}[\underline{\mathbf{u}}_1, \ldots, \underline{\mathbf{u}}_n]$. Note that each argument consists of three components, denoting a point in $\mathbb{P}^2$. If all n arguments agree, we obtain the point on the surface corresponding to that argument.

Now let $\underline{\mathbf{a}}, \underline{\mathbf{b}}, \underline{\mathbf{c}}$ be a second triangle in $\mathbb{P}^2$. Our polynomial $\underline{\mathbf{b}}(\mathbf{u})$ is defined over all of $\mathbb{P}^2$, and we may ask for its control points relative to the triangle $\underline{\mathbf{a}}, \underline{\mathbf{b}}, \underline{\mathbf{c}}$. They are simply given by

$$\underline{\mathbf{c}}_{\mathbf{i}} = \underline{\mathbf{b}}[\underline{\mathbf{a}}^{<i>}, \underline{\mathbf{b}}^{<j>}, \underline{\mathbf{c}}^{<k>}]; \quad |\mathbf{i}| = n. \tag{12.7}$$

Fig. 12.6 illustrates.[3]

A common application is the case of $\underline{\mathbf{a}}, \underline{\mathbf{b}}, \underline{\mathbf{c}}$ being the midpoint of the domain triangle's edges: then we split one triangular patch into four new ones, a process called *subdivision* just as in the curve case.

---

[3]This figure is the surface analog of Fig. 7.12.

We may also use blossoms to compute the control points of a curve on the surface: let $\underline{\mathbf{a}}$ and $\underline{\mathbf{b}}$ be two points in $I\!P^2$. They define a line, which will be mapped to a polynomial curve of degree n, lying completely on the surface $\underline{\mathbf{b}}(\underline{\mathbf{u}})$. The control points $\underline{\mathbf{p}}_i$ of this curve are simply

$$\underline{\mathbf{p}}_i = \underline{\mathbf{b}}[\underline{\mathbf{a}}^{<n-i>}, \underline{\mathbf{b}}^{<i>}]; \quad i = 0, \ldots, n. \tag{12.8}$$

Note the difference to rectangular patches: there a line in the domain mapped to a curve of degree $m + n$ in space!

Our projective reference frame for $I\!P^2$ implies the existence of a *fundamental line* $\underline{\mathbf{F}}$, see Section 1.5. Let $\underline{\mathbf{f}}$ be a point on it, whereas the vertices of the domain triangle are not on it. In complete analogy to our development of the curve case, we may use $\underline{\mathbf{f}}$ to compute *derivatives* of $\underline{\mathbf{b}}(\underline{\mathbf{u}})$ with respect to $\underline{\mathbf{f}}$:

$$D_{\underline{\mathbf{f}}}^r \underline{\mathbf{b}}(\underline{\mathbf{u}}) \doteq \underline{\mathbf{b}}[\underline{\mathbf{f}}^{<r>}, \underline{\mathbf{u}}^{<n-r>}]. \tag{12.9}$$

If we set $r = n$, then we see that the nth derivative of $\underline{\mathbf{b}}$ with respect to $\underline{\mathbf{f}}$ is on the image of $\underline{\mathbf{F}}$, which is a curve on the surface.

We shall consider the case $r = 1$ in more detail. Setting $\underline{\mathbf{f}} = [f, g, h]^{\mathrm{T}}$, we have

$$\begin{aligned} D_{\underline{\mathbf{f}}}^r \underline{\mathbf{b}}(\underline{\mathbf{u}}) & \doteq \underline{\mathbf{b}}[\underline{\mathbf{f}}^{<r>}, \underline{\mathbf{u}}^{<n-r>}] \\ & \doteq \sum_{|i|=n-1} [f\underline{\mathbf{b}}_{i+e1} + g\underline{\mathbf{b}}_{i+e2} + h\underline{\mathbf{b}}_{i+e3}] B_i^{n-1}(\underline{\mathbf{u}}). \end{aligned}$$

Thus the coefficients of the derivative surface (of degree $n - 1$) are formed by the terms $f\underline{\mathbf{b}}_{i+e1} + g\underline{\mathbf{b}}_{i|e2} + h\underline{\mathbf{b}}_{i+e3}$, obtained by application of one level of the de Casteljau algorithm to the original control points. Figure 12.7 illustrates.

If $\underline{\mathbf{f}}$ and $\underline{\mathbf{g}}$ are two points on $\underline{\mathbf{F}}$, we may define *mixed derivatives* in the same way:

$$D_{\underline{\mathbf{f}}, \underline{\mathbf{g}}}^{r,s} \underline{\mathbf{b}}(\underline{\mathbf{u}}) \doteq \underline{\mathbf{b}}[\underline{\mathbf{f}}^{<r>}, \underline{\mathbf{g}}^{<s>}, \underline{\mathbf{u}}^{<n-r-s>}]. \tag{12.10}$$

Note that it is not meaningful to consider expressions of the form $D_{\underline{\mathbf{f}}, \underline{\mathbf{g}}, \underline{\mathbf{h}}}^{r,s,t} \underline{\mathbf{b}}(\underline{\mathbf{u}})$ with all three points $\underline{\mathbf{f}}, \underline{\mathbf{g}}, \underline{\mathbf{h}}$ on $\underline{\mathbf{F}}$: this leads to a new domain triangle with three collinear vertices.

Let us investigate the relationship between blossoms and derivatives in more detail for the quadratic case $n = 2$. The fundamental line $\underline{\mathbf{F}}$ is mapped to a conic $\underline{\mathbf{c}}_f$ on the surface. Let $\underline{\mathbf{u}}$ be a point in the domain, and $\underline{\mathbf{b}}(\underline{\mathbf{u}})$ its corresponding surface point. With $\underline{\mathbf{f}}$ a point on $\underline{\mathbf{F}}$, the line $\underline{\mathbf{u}} \wedge \underline{\mathbf{f}}$ is mapped to a conic $\underline{\mathbf{c}}$ on the surface. The derivative of this conic at $\underline{\mathbf{u}}$ is

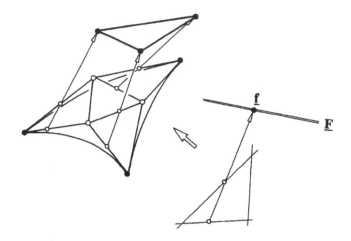

Figure 12.7. Derivatives: the derivative with respect to a point on the fundamental line in the domain is given by a surface whose control points are obtained by the blossoming principle.

the intersection of the conic's tangents at $\underline{\mathbf{b}}(\underline{\mathbf{u}})$ and at $\underline{\mathbf{b}}(\underline{\mathbf{f}})$.[4] Thus every derivative to $\underline{\mathbf{b}}$ lies on one of $\underline{\mathbf{b}}$'s tangent planes at $\underline{\mathbf{c}}_f$. Note that these tangent planes generate a developable surface, see Section 11.10.

In the degenerate case where $\underline{\mathbf{c}}_f$ collapses to a single point $\underline{\mathbf{p}}$, all tangent planes along $\underline{\mathbf{c}}$ degenerate to the tangent plane at $\underline{\mathbf{p}}$, and all first derivatives of $\underline{\mathbf{b}}$ lie in this plane. Quadratic surfaces for which $\underline{\mathbf{F}}$ is mapped to a single point $\underline{\mathbf{p}}$ are obtainable by a *stereographic projection* through $\underline{\mathbf{p}}$, see Section 13.3. In this special case, the quadratic surface is actually part of a *quadric*; see [25], [146], [110].

12.4   Barycentric coordinates

In the projective plane, any four points $\underline{\mathbf{a}}, \underline{\mathbf{b}}, \underline{\mathbf{c}}, \underline{\mathbf{d}}$ define a coordinate system by setting $\underline{\mathbf{d}} = \underline{\mathbf{a}} + \underline{\mathbf{b}} + \underline{\mathbf{c}}$. We can then write any other point $\underline{\mathbf{x}}$ in $I\!\!P^2$ as $\underline{\mathbf{x}} = u\underline{\mathbf{a}} + v\underline{\mathbf{b}} + w\underline{\mathbf{c}}$. This representation is unique up to a common factor in the $\underline{\mathbf{x}}$'s coordinates u, v, w. See Section 1.5 for more details.

Since $(u, v, w) = (0, 0, 0)$ does not denote a point location, we may normalize our projective coordinates such that

$$u + v + w = 1. \tag{12.11}$$

---

[4]Recall that the derivative of a conic at a given point is the intersection with the tangent at that point with the tangent at $t = \infty$.

Figure 12.8. Barycentric coordinate systems: the three families of isoparametric lines are shown.

Projective coordinates that satisfy (12.11) are called *barycentric coordinates* after F. Moebius [108]. Their main importance is not in projective, but in affine geometry.

If we project the relation $\underline{\mathbf{d}} = \underline{\mathbf{a}} + \underline{\mathbf{b}} + \underline{\mathbf{c}}$ into affine space, we obtain

$$\mathbf{d} = \frac{1}{3}\mathbf{a} + \frac{1}{3}\mathbf{b} + \frac{1}{3}\mathbf{c},$$

as was outlined in Section 1.10. This has the geometric interpretation that $\mathbf{d}$ is the centroid of the triangle $\underline{\mathbf{a}}, \underline{\mathbf{b}}, \underline{\mathbf{c}}$.

Just as in the projective plane, we write every point $\mathbf{x}$ of the extended affine plane as a *barycentric combination*.

$$\mathbf{x} = u\mathbf{a} + v\mathbf{b} + w\mathbf{c}; \quad u + v + w = 1.$$

Fig. 12.8 shows the three families of constant parameter values, in complete analogy to Fig. 1.9. Note that now the three families of constant parameter values each contain parallel lines. Thus the centers of the corresponding pencils all lie on the line at infinity, and hence are collinear. Barycentric coordinates are special projective coordinates: the fundamental line (see Section 1.5) is the line at infinity.

If a point $\mathbf{p}$ is given and we seek its barycentric coordinates $\mathbf{u} = (u, v, w)$ with respect to an affine triangle $\mathbf{a}, \mathbf{b}, \mathbf{c}$, we find

$$u = \frac{\text{area}(\mathbf{p}, \mathbf{b}, \mathbf{c})}{\text{area}(\mathbf{a}, \mathbf{b}, \mathbf{c})}, \quad v = \frac{\text{area}(\mathbf{a}, \mathbf{p}, \mathbf{c})}{\text{area}(\mathbf{a}, \mathbf{b}, \mathbf{c})}, \quad w = \frac{\text{area}(\mathbf{a}, \mathbf{b}, \mathbf{p})}{\text{area}(\mathbf{a}, \mathbf{b}, \mathbf{c})}, \tag{12.12}$$

with the understanding that the area of a triangle has a sign! The area of an affine triangle is given by $\text{area}(\mathbf{a}, \mathbf{b}, \mathbf{c}) = \det(\mathbf{a} - \mathbf{c}, \mathbf{b} - \mathbf{c})/2$. Here, each point has *two* coordinates!

12.5   Rational Bézier triangles

A rational Bézier triangle is the projection of a Bézier triangle in $I\!\!P^3$ into $I\!\!E^3$. We thus have:

$$\mathbf{b}^n(\mathbf{u}) = \mathbf{b}_0^n(\mathbf{u}) = \frac{\sum_{|\mathbf{i}|=n} w_{\mathbf{i}} \mathbf{b}_{\mathbf{i}} B_{\mathbf{i}}^n(\mathbf{u})}{\sum_{|\mathbf{i}|=n} w_{\mathbf{i}} B_{\mathbf{i}}^n(\mathbf{u})}, \tag{12.13}$$

where, as usual, the $w_{\mathbf{i}}$ are the weights associated with the control vertices $\mathbf{b}_{\mathbf{i}}$. Needless to say, for positive weights we have the convex hull property, and we have affine and projective invariance.

Rational Bézier triangles may be evaluated by a de Casteljau algorithm in a not too surprising way:

Rational de Casteljau algorithm

Given: a triangular array of points $\mathbf{b}_{\mathbf{i}} \in I\!\!E^3$; $|\mathbf{i}| = n$, corresponding weights $w_{\mathbf{i}}$, and a point in a domain triangle with barycentric coordinates $\mathbf{u}$.

Set:

$$\mathbf{b}_{\mathbf{i}}^r(\mathbf{u}) = \frac{u w_{\mathbf{i}+\mathbf{e1}}^{r-1} \mathbf{b}_{\mathbf{i}+\mathbf{e1}}^{r-1} + v w_{\mathbf{i}+\mathbf{e2}}^{r-1} \mathbf{b}_{\mathbf{i}+\mathbf{e2}}^{r-1} + w w_{\mathbf{i}+\mathbf{e3}}^{r-1} \mathbf{b}_{\mathbf{i}+\mathbf{e3}}^{r-1}}{w_{\mathbf{i}}^r} \tag{12.14}$$

where

$$w_{\mathbf{i}}^r = w_{\mathbf{i}}^r(\mathbf{u}) = u w_{\mathbf{i}+\mathbf{e1}}^{r-1}(\mathbf{u}) + v w_{\mathbf{i}+\mathbf{e2}}^{r-1}(\mathbf{u}) + w w_{\mathbf{i}+\mathbf{e3}}^{r-1}(\mathbf{u})$$

and

$$r = 1, \ldots, n \quad \text{and} \quad |\mathbf{i}| = n - r$$

and $\mathbf{b}_{\mathbf{i}}^0(\mathbf{u}) = \mathbf{b}_{\mathbf{i}}, w_{\mathbf{i}}^0 = w_{\mathbf{i}}$. Then $\mathbf{b}_0^n(\mathbf{u})$ is the point with parameter value $\mathbf{u}$ on the rational Bézier triangle $\mathbf{b}^n$.

This algorithm can be interpreted as projecting every intermediate point $\underline{\mathbf{b}}_i^r$ into affine space. Of course, we could evaluate all four components of $\underline{\mathbf{b}}(\mathbf{u})$ separately and divide through by the $w-$ component as a final step.

Here is a formula for the *directional derivative* of a rational Bézier triangle. Let $\mathbf{d}$ denote a direction in the domain triangle, expressed in barycentric coordinates. This means that $\mathbf{d}$ is the difference between two points, and thus a vector. Note that $\mathbf{d}$'s barycentric coordinates sum to zero. We are interested in the directional derivative $D_{\mathbf{d}}$ of a rational triangular Bézier patch $\mathbf{b}^n(\mathbf{u})$. Proceeding exactly as in the curve case, eq. (7.28), we obtain:

$$D_{\mathbf{d}}\mathbf{b}^n(\mathbf{u}) = \frac{1}{w(\mathbf{u})}[\dot{\mathbf{p}}(\mathbf{u}) - D_{\mathbf{d}}(\mathbf{u})\mathbf{b}^n(\mathbf{u})],$$

where we have set

$$\mathbf{p}(\mathbf{u}) = w(\mathbf{u})\mathbf{b}^n(\mathbf{u}) = \sum_{|\mathbf{i}|=n} w_{\mathbf{i}}\mathbf{b}_{\mathbf{i}}B_{\mathbf{i}}^n(\mathbf{u}).$$

Higher derivatives follow the pattern of (7.31):

$$D_{\mathbf{d}}^r\mathbf{b}^n(\mathbf{u}) = \frac{1}{w(\mathbf{u})}\left[D_{\mathbf{d}}^r\mathbf{p}(\mathbf{u}) - \sum_{j=1}^{r}[D_{\mathbf{d}}^j w(\mathbf{u})][D_{\mathbf{d}}^{r-j}(\mathbf{u})]\right].$$

Similar to the rectangular case in Section 11.3, we may use the intermediate points in the de Casteljau algorithm to determine the *tangent plane* corresponding to a parameter value $\mathbf{u}$: it is given by the three $\mathbf{b}_{\mathbf{e1}}^{n-1}(\mathbf{u}), \mathbf{b}_{\mathbf{e2}}^{n-1}(\mathbf{u}), \mathbf{b}_{\mathbf{e3}}^{n-1}(\mathbf{u})$.

An example of a rational Bezier triangle is given in Section 15.6; there, an octant of a sphere is described.

12.6   Problems

1. The isoparametric lines $u = const, v = const, w = const$ of a rational quadratic Bézier triangle are conics. At a given point, we can thus define three conics through it. Can all of these conics be of the same type?

2. What is the minimum degree for a rational Bézier triangle such that it has *two* ellipses as the intersection with a plane?

3. What is the minimum degree for a rational Bézier triangle such that it can define a surface with the topology of a torus?

4. Let $\underline{u}(t)$ be the parametric form of a quadratic Bézier curve in a Bézier triangle's domain. What are the Bézier points of the line segment corresponding to $[\underline{u}(0), \underline{u}(1)]$?

13

Quadrics

The most fundamental curves in the projective plane are the conics. As it turned out, they can be represented by projective quadratic Bézier curves. Conversely, every projective Bézier curve represents a conic.

For surfaces, life is not that easy. The fundamental surfaces in projective space are the *quadrics* — but there is no one-to-one correspondence between them and any kind of Bézier surface. In this chapter, we shall define quadrics, and then explore the relationship between quadrics and Bézier surfaces.

13.1   Quadrics

A quadric in $I\!\!P^3$ is a surface that has an implicit equation of the form

$$\underline{x}^T A \underline{x} = 0, \tag{13.1}$$

with $\underline{x} = [x, y, z, w]^T$. A quadric has no more than two intersections with any straight line. Any planar section of a quadric is a conic. In fact, an alternative definition of a quadric is as being a surface all of whose planar sections are conics.

Figure 13.1. Tangent planes to a quadric: the shown configuration must satisfy Brianchon's theorem.

A quadric is defined by nine points on it. The implicit form takes on the form

$$
\begin{vmatrix}
x^2 & y^2 & z^2 & w^2 & xy & xz & yz & xw & yw & zw \\
x_1^2 & y_1^2 & z_1^2 & w_1^2 & x_1y_1 & x_1z_1 & y_1z_1 & x_1w_1 & y_1w_1 & z_1w_1 \\
\vdots & \vdots & \vdots & \vdots & \vdots & \vdots & \vdots & \vdots & \vdots & \vdots \\
x_9^2 & y_9^2 & z_9^2 & w_9^2 & x_9y_9 & x_9z_9 & y_9z_9 & x_9w_9 & y_9w_9 & z_9w_9
\end{vmatrix} = 0,
$$

(13.2)

much in the same way as (4.26) described a conic through five points.

Let $\underline{a}, \underline{b}, \underline{c}$ be any three noncollinear points on a quadric. They form a plane $\underline{P}$, which cuts a conic Γ out of the quadric. The tangent planes at $\underline{a}, \underline{b}, \underline{c}$ cut tangents to Γ out of $\underline{P}$, see Fig. 13.1. The three points $\underline{a}, \underline{b}, \underline{c}$ together with the tangents, all belonging to Γ, must then satisfy Brianchon's theorem (see Fig. 3.14): the three heavy lines in Fig. 13.1 are concurrent. This shows that three tangent planes to a quadric are not independent of each other. In particular, a quadric is not determined by three points and three tangent planes, although these might naively be counted as nine points.

13.2 Quadric classification

All quadrics in $I\!P^3$ can be classified into three categories:[1]

i) If for every point on the quadric, the tangent plane intersects the quadric in two distinct straight lines, the quadric is called doubly ruled or annular,

ii) If for every point on the quadric, the tangent plane intersects the quadric in exactly one straight line, the quadric is called singly ruled or a cone,

---

[1]We limit ourselves to *real* quadrics here — if we allow *complex* quadrics, the distinction between the three quadric types disappears.

Figure 13.2. Quadric categories: i) doubly ruled, ii) cone, iii) oval.

iii) If for every point on the quadric, the tangent plane does not intersect the quadric in any straight line, the quadric is called oval.

Since straight lines are mapped to straight lines by projective maps, it follows that quadrics can only be mapped to quadrics in the same category. Fig. 13.2 shows the local geometry of the three categories. This is in sharp contrast to the curve case: there, every conic could be obtained as a projective map of any other conic. Now, we have to pay attention to the kind of quadric that we are dealing with.

Affine quadrics are obtained by projecting the projective quadrics into affine space. Obviously, the category of a quadric is not changed by this projection. When we discussed affine conics, we studied their behavior at the line at infinity. Consequently, we are now interested in the behavior of quadrics at the plane at infinity. Let $\underline{\mathbf{H}}$ be the projective plane that will be mapped to the affine plane at infinity. We will call affine quadrics

- *hyperboloids* if they intersect $\underline{\mathbf{H}}$,

- *ellipsoids* have no point in common with $\underline{\mathbf{H}}$,

- *paraboloids* if they are tangent to $\underline{\mathbf{H}}$,

in complete analogy to the conic classification from Section 3.5. If the projective quadric intersects $\underline{\mathbf{H}}$, the resulting conic will be mapped to an affine *conic at infinity.*

Let us now discuss each of the three categories.

i) We know that a *doubly ruled* projective quadric either intersects a given plane, or it is tangent to it. This is also true for $\underline{\mathbf{H}}$. In the case of an intersection, the resulting affine quadric will be a hyperboloid, called *hyperboloid of one sheet.* In the case of tangency, we obtain a paraboloid, namely the *hyperbolic paraboloid.*

ii) A singly ruled quadric may be of three types, depending on the type of the conic at infinity. We distinguish between hyperbolic, parabolic,

Figure 13.3. Affine quadrics: top row, elliptic paraboloid, ellipsoid, hyperboloid of two sheets; middle row: hyperbolic paraboloid, hyperboloid of one sheet; bottom row: parabolic cylinder, elliptic cylinder, hyperbolic cylinder, cone.

and elliptic cones. In the special case that all straight lines on the cone are parallel, we have hyperbolic, parabolic, and elliptic *cylinders*. The parabolic cylinder really is a paraboloid, touching the plane at infinity along a straight line.

iii) A projective *oval* quadric may intersect **H** in a conic; then the corresponding affine quadric is a hyperboloid and is called *hyperboloid of two sheets*. If the projective oval quadric touches **H**, the resulting affine conic is a paraboloid; it is called *elliptic paraboloid*. If the projective oval quadric has no point in common with **H**, the resulting affine quadric is simply called an *ellipsoid*. It is the only finite affine quadric.

One can list seventeen types of affine quadrics — but for that, one has to count all kinds of degeneracies. For example, a pair of parallel planes and a pair of intersecting planes constitute two more examples of quadrics. For details, see the literature on projective and / or analytic geometry, for example [18], [26], [86], [144]. In affine n−space, there are 42 different kinds of quadrics, see [117][2] We illustrate the nondegenerate 3D affine quadrics in Fig. 13.3.

---

[2] Also recall the final answer in the "hitchhiker's guide to the galaxy!"

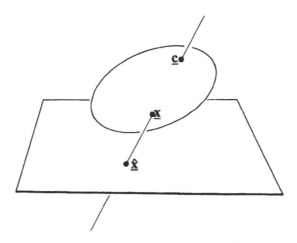

Figure 13.4. Stereographic projections: a point on a quadric is projected onto a plane, with the center of the projection being on the quadric.

13.3 The stereographic projection

We are interested in the relationship between Bézier patches and quadrics; in order to address that problem, we need a tool: the *stereographic projection*. Let $\underline{\mathbf{Q}}$ denote a quadric surface, and let $\underline{c}$ be a point on it, and further let $\underline{\mathbf{P}}$ be an arbitrary plane not through $\underline{c}$. See Fig. 13.4.

If $\underline{x}$ is an arbitrary point on the quadric, we may form the line through $\underline{c}$ and $\underline{x}$ and intersect it with the plane $\underline{\mathbf{P}}$, resulting in a point $\hat{\underline{x}}$. We say that $\hat{\underline{x}}$ is the image of $\underline{x}$ under the stereographic projection with center $\underline{c}$.

There are two important geometric properties of stereographic maps:

1. They provide a one-to-one map between the quadric and the plane, except for one singular point.

2. The image of a conic on the quadric is a conic on the plane.

The most important *algebraic* property of stereographic maps is that they can be described by quadratic maps. To make this notion more precise, we follow a development by Teller and Séquin [146]. Referring to Fig. 13.5, let $\underline{b}_{002}, \underline{b}_{020}$, and $\underline{b}_{200}$ be three points on the quadric. We choose this naming convention as we shall make these three points the corners of a triangular quadratic Bézier patch. The stereographic projection of these three points yields a triangle in the plane, which will serve as the domain for the triangular patch.

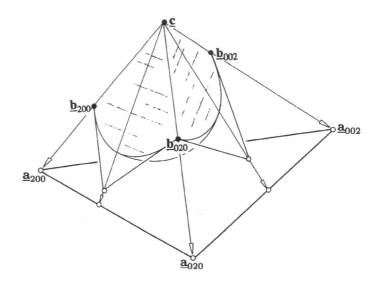

Figure 13.5. Stereographic projections: a quadratic Bézier triangle can be utilized to describe the map.

Our desired Bézier patch will have three planar boundary curves, and for these we choose the planes through $\underline{c}$ and the three corner points $\underline{b}_{002}, \underline{b}_{020}$, and $\underline{b}_{200}$. Each of these planes cuts a conic out of the quadric, and these conics will be our patch boundary curves. Let us concentrate on the boundary curve $v = 0$, lying in the plane spanned by $\underline{c}, \underline{b}_{002}, \underline{b}_{200}$. It will be of the form

$$\underline{b}(t) = \underline{b}_{002} B_0^2(t) + \underline{b}_{101} B_1^2(t) + \underline{b}_{200} B_2^2(t). \tag{13.3}$$

We can find the Bézier point $\mathbf{b}_{101}$ by intersecting the plane of the boundary curve with the tangent planes at $\underline{b}_{002}$ and $\underline{b}_{200}$, see Fig. 13.6. We pick the shoulder tangent such that $\underline{c}$ corresponds to $t = \infty$.

The remaining boundary curves should have the same form as (13.3). This is achieved by also selecting $\underline{c}$ as corresponding to the parameter value ∞ for all boundary curves.

We now have completed the construction of a quadratic patch, interpolating to data from the quadric. In fact, this patch *is* the quadric since it agrees with the quadric in three points and the tangent planes there, plus in another point $\underline{c}$.

In our development, we have not only constructed a patch on the quadric, but we have also fixed a coordinate system on the plane: let $\underline{a}_{ijk}$

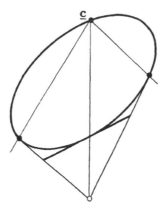

Figure 13.6. Stereographic projections: a boundary curve is determined by its Bézier points and a shoulder tangent.

be the projection of $\underline{\mathbf{b}}_{ijk}$ through $\underline{\mathbf{c}}$ onto the plane. Then the three lines $\mathbf{a}_{002} \wedge \mathbf{a}_{110}$, $\mathbf{a}_{020} \wedge \mathbf{a}_{101}$, and $\mathbf{a}_{200} \wedge \mathbf{a}_{011}$ are concurrent, their intersection $\mathbf{a}_{111}$ being the projection of the point $\underline{\mathbf{b}}(\frac{1}{3}, \frac{1}{3}, \frac{1}{3})$. The four points $\mathbf{a}_{002}, \mathbf{a}_{020}, \mathbf{a}_{200}$, and $\mathbf{a}_{111}$ thus form a projective reference frame for the plane: every point $\underline{\mathbf{a}}$ in the plane can be written as $\underline{\mathbf{a}} = u\mathbf{a}_{200} + v\mathbf{a}_{020} + w\mathbf{a}_{002}$. The point $\underline{\mathbf{b}}(u, v, w)$ is then the image of $\underline{\mathbf{a}}$.

To summarize: starting with a quadric, a point on it, and a plane, we defined a stereographic projection. We have then shown that this stereographic projection can be interpreted as a quadratic map of the plane to the quadric, i.e., every point $\underline{\mathbf{x}}$ on the quadric can be written as

$$\underline{\mathbf{x}} = \underline{\mathbf{x}}(u, v, w) = \sum_{|i|=2} \mathbf{b}_i B_i^2(u, v, w).$$

Note that this quadratic map is not unique: it depends on the choice of the three patch corners.

If we define a Bézier curve in the plane (with the reference frame that we just defined), then it is of the form

$$\underline{\mathbf{u}}(t) = \sum_{i=0}^{n} \underline{\mathbf{c}}_i B_i^n(t); \quad \underline{\mathbf{x}}, \underline{\mathbf{c}}_i \in I\!P^2. \tag{13.4}$$

Using the stereographic projection, we can map it to a curve on the quadric, where it will be a Bézier curve of degree $2n$. What are its Bézier points $\underline{\mathbf{o}}_i$? The answer is rather straightforward when we employ the blossom of the quadric.

Recall that the blossom is defined by $\underline{b}[\underline{u}, \underline{v}]$ by simply using two different arguments $\underline{u}$ and $\underline{v}$ in the de Casteljau algorithm. A point on the quadric is obtained if the two arguments are identical. See Section 12.3 for more details.

For any parameter value t, we obtain a point $\underline{u}(t)$ in the plane and thus a point $\underline{x}(t) = \underline{b}[\underline{u}(t), \underline{u}(t)]$ on the quadric. This leads to the following development (making use of the multilinearity properties of blossoms):

$$
\begin{aligned}
\underline{b}[\underline{u}(t), \underline{u}(t)] &= \underline{b}[\sum_{i=0}^{n} \underline{c}_i B_i^n(t), \sum_{j=0}^{n} \underline{c}_j B_j^n(t)] \\
&= \sum_{i=0}^{n} \underline{b}[\underline{c}_i B_i^n(t), \sum_{j=0}^{n} \underline{c}_j B_j^n(t)] \\
&= \sum_{i=0}^{n} \sum_{j=0}^{n} \underline{b}[\underline{c}_i B_i^n(t), \underline{c}_j B_j^n(t)] \\
&= \sum_{i=0}^{n} \sum_{j=0}^{n} \underline{b}[\underline{c}_i, \underline{c}_j] B_i^n(t) B_j^n(t).
\end{aligned}
$$

Using the simple fact that

$$
B_i^n(t) B_j^n(t) = \frac{\binom{n}{i}\binom{n}{j}}{\binom{2n}{i+j}} B_{i+j}^{2n}(t),
$$

we can now write

$$
\underline{x}(t) = \sum_{l=0}^{2n} \underline{o}_l B_l^{2n}(t) \tag{13.5}
$$

with

$$
\underline{o}_l = \sum_{i+j=l} \frac{\binom{n}{i}\binom{n}{j}}{\binom{2n}{i+j}} \underline{b}[\underline{c}_i, \underline{c}_j]. \tag{13.6}
$$

Example 13.1 illustrates for the "standard" stereographic projection onto the unit sphere through its north pole. We have seen this before: our special stereographic projection is nothing but a rewrite of (10.20)!

It also possible to compute the $\underline{o}_l$ if the quadric is given in implicit form:

$$
\underline{o}_l = \sum_{i+j=l} \frac{\binom{n}{i}\binom{n}{j}}{\binom{2n}{i+j}} [\underline{c}(\underline{c}_i^T A \underline{c}_j) - 2\underline{c}_i(\underline{c}^T A \underline{c}_j)], \tag{13.7}
$$

quoting from Boehm [24]. Again, $\underline{c}$ is the center of the projection.

The stereographic map (10.20) may be written as a homogeneous quadratic Bézier patch. It coefficients are given by

$$\begin{bmatrix} 0 \\ 0 \\ 1 \\ 1 \end{bmatrix}$$

$$\begin{bmatrix} 0 \\ 1 \\ 0 \\ 0 \end{bmatrix} \begin{bmatrix} 1 \\ 0 \\ 0 \\ 0 \end{bmatrix}$$

$$\begin{bmatrix} 0 \\ 0 \\ -1 \\ 1 \end{bmatrix} w \begin{bmatrix} 0 \\ 0 \\ 0 \\ 0 \end{bmatrix} \begin{bmatrix} 0 \\ 0 \\ 1 \\ 1 \end{bmatrix}$$

Example 13.1. A special stereographic projection.

Let us conclude this section with some further (projective) analysis of a quadratic patch on a quadric. The projective reference frame in the plane defines a fundamental line $\mathbf{F}$ in the plane, as illustrated earlier in Fig. 1.9. It contains the points $(-\infty, \infty, 0), (-\infty, 0, \infty)$, and $(0, -\infty, \infty)$. Now recall that the point $\underline{c}$ on the quadric corresponded to just those values, for that was how we parametrized the patch boundary curves. It follows that the whole fundamental line is mapped to $\underline{c}$ by the stereographic projection! In other words, the stereographic projection is not one to one for $\mathbf{F}$.

It then follows that the plane defined by $\underline{c}$ and $\mathbf{F}$ is the quadric's tangent plane at $\underline{c}$; see Fig. 13.7 for an illustration of the case of a cone.

Quadratic surfaces have been studied in detail by J. Steiner; thus one also encounters the name "Steiner patches" for quadratic Bézier triangles, see Sederberg [137].

13.4 Quartics on quadrics

If we are given a quadratic curve in a plane — let us take $I\!\!P^2$ for simplicity — and a quadric with a point $\underline{c}$ on it, what is the image of the given quadratic under the stereographic projection through $\underline{c}$? Given that the stereographic projection is a quadratic map, we would expect the image to be of degree four. The geometry is illustrated in Fig. 13.8; the formulas follow exactly the development in Section 10.5.

Figure 13.7. Stereographic projections: the tangent plane at the center intersects the projection plane in the fundamental line.

We may next ask: given a quadric in $I\!P^3$, a point on it, and a plane,
we could find quadratic patches on the quadric by means of stereographic
projections. If we do not refer to the stereographic projection, we may
describe the resulting quadratic patches as follows:

1. The three boundary curves meet in one point $\underline{c}$.

2. The three boundary curves are coplanar at $\underline{c}$.

3. Each boundary curve assumes the parameter value $\pm\infty$ at $\underline{c}$.

Figure 13.8. Curves on quadrics: a quadratic Bézier curve in the plane is
mapped to a degree four Bézier curve on the quadric.

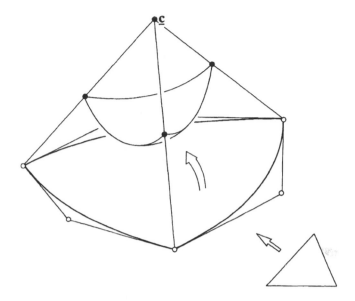

Figure 13.9. Nonquadratic patches on quadrics: the three boundary curves may be projected onto three conics in a plane.

So if we encounter a quadratic patch that satisfies the above three conditions, we may conclude that it is a patch on a quadric: the information provided is sufficient to uniquely determine a quadric. A warning is in place: the above conditions cannot be checked in a numerically reliable way for many cases!

Not all quadratic patches are on quadrics, since not all quadratic patches meet the three conditions. In terms of algebraic geometry, a quadratic patch is said to have *algebraic degree* four, meaning that a quadratic patch and a straight line may have up to four intersections.[3] A quadric, on the other hand, can only have two such intersections.

If we are given three planar sections of a quadric (remember: planar sections of quadrics are always conics!) as shown in Fig. 13.9, can we find a triangular patch with those conics as boundaries that is on the given quadric? The answer is yes, but the patch will not be of degree two — unless, of course, the above three conditions are met.

The stereographic projection may again be used to find the desired patch formulation. Pick any point $\underline{c}$ on the quadric, and an arbitrary plane not through $\underline{c}$. Project the three boundary curves, together with their

---

[3] Counting coincident as well as complex intersections.

Figure 13.10. Quadrics: an ellipsoid covered with quartic patches.

control polygons, into the plane, using a projection with center $\underline{c}$. The images of the conics will be three conics, as shown in Fig. 13.9. Let the plane have projective coordinates u, v, w.

The three conics may be interpreted as the boundary curves of a planar quadratic patch $\underline{a}$. It is the image of a triangle with projective coordinates (r, s, t):

$$\underline{a}(r, s, t) = \sum_{|\mathbf{i}|=2} \underline{a}_\mathbf{i} B_\mathbf{i}^2(r, s, t).$$

This quadratic Bézier triangle is a map of the (r, s, t) plane onto the (u, v, w) plane. We know that the stereographic projection is a quadratic map (once we picked three points on the quadric and proceeded as in Section 13.3.) Thus our triangular patch on the quadric is the composition of the two quadratic maps, making it a degree four patch, a quartic.[4]

There is one important difference between this kind of quartic patch versus the quadratic one discussed above: the quadratic patch covers the quadric completely, and, with the exception of $\underline{c}$, provides a one-to-one coverage. The quartic, on the other hand, does neither: it does not cover the whole quadric; some parts of the quadric are covered twice, others not at all. The reason is that the patch $\underline{u} = \underline{a}(r, s, t)$ does not cover all of $\mathbb{P}^2$; it may be interpreted as a "flattened out" quadric itself. This phenomenon was observed by M. Niebuhr [110].

---

[4]To avoid a terminology overload, the basic definitions at a glance: quadratic: being of degree two; quartic: being of degree four; quadric: a surface with algebraic degree two.

Figure 13.11. Quadrics: a hyperbolic paraboloid covered with quartic patches.

Any triangular patch on a quadric can be written as a degree four Bézier triangle, provided it has planar boundary curves. Once projected into affine space, the quartic patch becomes a rational quartic, and so all triangular patches (with planar boundary curves) on affine quadrics can be written as rational quartic Bézier triangles. In the special case of a sphere, Section 15.6 gives a numerical example. Some illustrations are provided by Figures 13.10, 13.11, and 13.12.

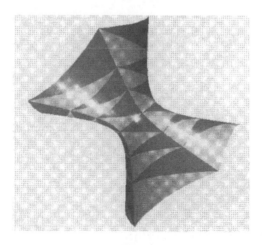

Figure 13.12. Quadrics: a hyperboloid of one sheet covered with quartic patches.

In the same manner, one may obtain a rectangular biquartic patch on the sphere, following J. Cobb. Its Bézier points and their weights are listed in Section 15.7. Six such patches (suitably rotated) cover the sphere with a rectangular network that has a cube topology.

For more literature regarding rational patches on quadrics, see also [90], [88], [45].

13.5   Problems

1. If a rational cubic triangular patch (with its domain extended to be the whole extended affine plane) is *closed*, does this mean it is a quadric?

2. Find a rational quadratic patch that has four intersections with some straight line.

3. A triangle divides the projective plane into four regions, see Section 1.11. Mark these regions on a quadric that is defined by a quadratic Bézier triangle.

4. Given a rational quadratic patch on a quadric (i.e., one that satisfies the three conditions from Section 13.4), devise a method to decide its type — doubly ruled, singly ruled, or oval. See [7].

5. What is the geometric meaning of the term $\underline{\mathbf{b}}[\underline{\mathbf{c}}_i, \underline{\mathbf{c}}_j]$ in Eq. (13.6)?

14

Gregory Patches

Bicubic patches (integral ones) have enjoyed enormous popularity. Another type of surface, the Coons patch, never became quite as popular because of problems people had with a correct specification of certain twist vectors (see [57] for details.) The twist problem was first fixed by J. Gregory in 1972, still in the context of Coons patches. What really made his idea recognized and successful was, however, a modification by Chiyokura and Kimura as described in [33] and [32]. The following sections will focus on that approach; for the original literature, see Barnhill [14], Barnhill and Gregory [15], and Gregory [83].[1]

J. Gregory has contributed more to the field of Geometric Modeling than just the patches that now bear his name. He also obtained fundamental results in the areas of interpolatory subdivision, n–sided patches, and shape preserving interpolation. John Gregory died in March 1993, at the age of 47.

14.1   Bicubic Gregory patches

Consider the problem that is illustrated by Fig. 14.1: two (integral) bicubic Bézier patches A and B are given. A third one is sought such that we

---

[1]In the earlier literature, Gregory patches are typically referred to as "Gregory squares".

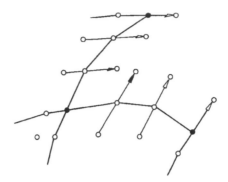

Figure 14.1. Gregory patches: the given data (solid circles) are incompatible.

achieve C^1 continuity across all patch boundaries.

The C^1 conditions between rectangular Bézier patches determine all points in the desired patch — with one exception: the pair of points marked by squares. Of those two, the point $\mathbf{b}_{1,1}^{(u)}$ is determined by patch A, while $\mathbf{b}_{1,1}^{(v)}$ is determined by patch B. Obviously, no bicubic Bézier patch can have two different points $\mathbf{b}_{1,1}$, and so the given problem has no solution. The way out of this dilemma is the inception of a rational patch that *does* incorporate two different points $\mathbf{b}_{1,1}$, namely both $\mathbf{b}_{1,1}^{(u)}$ and $\mathbf{b}_{1,1}^{(v)}$.

The key idea is to make $\mathbf{b}_{1,1}$ a function of u and v:

$$\mathbf{b}_{11} = \mathbf{b}_{11}(u,v) = \frac{u\mathbf{b}_{11}^{(v)} + v\mathbf{b}_{11}^{(u)}}{u+v}.$$

We also define

$$\mathbf{b}_{21} = \frac{(1-u)\mathbf{b}_{21}^{(v)} + v\mathbf{b}_{21}^{(u)}}{(1-u)+v},$$

$$\mathbf{b}_{12} = \frac{u\mathbf{b}_{12}^{(v)} + (1-v)\mathbf{b}_{12}^{(u)}}{u+(1-v)},$$

$$\mathbf{b}_{22} = \frac{(1-u)\mathbf{b}_{22}^{(v)} + (1-v)\mathbf{b}_{22}^{(u)}}{(1-u)+(1-v)}.$$

This is the Chiyokura/Kimura approach, based on the original approach by J. Gregory. It replaces the terms $\mathbf{b}_{11}$ etc. in the original Bézier formulation by the above terms. Note that if $\mathbf{b}_{11}^{(u)} = \mathbf{b}_{11}^{(v)}$ etc., the bicubic Gregory patch becomes a standard bicubic Bézier patch.

Figure 14.2. Gregory patches: the cross-boundary derivative along $u = 0$ only depends on the indicated difference vectors.

What have we gained? Let us take $v-$ partials of the Gregory patch along the boundary curve $v = 0$. We obtain

$$
\begin{aligned}
\mathbf{x}_v(u, 0) \;=\;& 3 * \big[(1 - u)^3[\mathbf{b}_{01} - \mathbf{b}_{00}] + 3(1 - u)^2 u[\mathbf{b}_{11}^{(u)} - \mathbf{b}_{10}] \\
& + 3(1 - u)u^2[\mathbf{b}_{21}^{(u)} - \mathbf{b}_{20}] + u^3[\mathbf{b}_{31} - \mathbf{b}_{30}]\big].
\end{aligned}
\tag{14.1}
$$

In words: the cross-boundary derivative $\mathbf{x}_v(u, 0)$ only depends on the control points as shown in Fig. 14.2. Thus it solves the incompatibility problem highlighted in Fig. 14.1, since analogous statements can be made for the remaining boundary curves.

It is now time to describe another important property of Gregory patches their corner twists[2] are *discontinuous*. To make this claim more precise, consider the corner twist $\mathbf{x}_{uv}(0, 0)$. When we try to compute it, we end up with an expression of the form $\mathbf{0}/0$, a meaningless expression. But we may evaluate $\mathbf{x}_{uv}(0, 0)$ by taking certain limits. If we approach $(0, 0)$ along $u = 0$, we obtain

$$
\lim_{u \to 0} \mathbf{x}_{uv}(0, 0) = 9[\mathbf{b}_{11}^u - \mathbf{b}_{10} - \mathbf{b}_{01} + \mathbf{b}_{00}].
$$

Taking a different limit, by approaching $(0, 0)$ along $v = 0$, we obtain

$$
\lim_{v \to 0} \mathbf{x}_{vu}(0, 0) = 9[\mathbf{b}_{11}^v - \mathbf{b}_{10} - \mathbf{b}_{01} + \mathbf{b}_{00}].
$$

14.2   The rational Bézier form

The appearance of terms such as $u/(u + v)$ in our Gregory patches make them *rational*. However, they are not of the form (11.3). But after some

---

[2]These are the patches' mixed partials, evaluated at the corners.

rewriting, one obtains a rational Bézier patch of degree seven in both u and v. Quoting [145], the Bézier points $\mathbf{c}_{ij}$:

$$\mathbf{c}_{00} = \mathbf{b}_{00},$$
$$\mathbf{c}_{01} = \mathbf{b}_{00},$$
$$\mathbf{c}_{02} = (5\mathbf{b}_{00} + 6\mathbf{b}_{01})/11,$$
$$\mathbf{c}_{03} = (2\mathbf{b}_{00} + 15\mathbf{b}_{01} + 6\mathbf{b}_{02})/23,$$
$$\mathbf{c}_{11} = (7\mathbf{b}_{00} + 3\mathbf{b}_{01} + 3\mathbf{b}_{01})/13,$$
$$\mathbf{c}_{12} = (18\mathbf{b}_{11}^{(v)} + 28\mathbf{b}_{00} + 42\mathbf{b}_{01} + 6\mathbf{b}_{02} + 15\mathbf{b}_{10})/109,$$
$$\mathbf{c}_{13} = (45\mathbf{b}_{11}^{(v)} + 18\mathbf{b}_{12}^{(u)} + 14\mathbf{b}_{00} + 84\mathbf{b}_{01} + 42\mathbf{b}_{02} + 2\mathbf{b}_{03} + 6\mathbf{b}_{10})/211$$
$$\mathbf{c}_{22} = (63\mathbf{b}_{11}^{(u)} + 63\mathbf{b}_{11}^{(v)} + 18\mathbf{b}_{12}^{(v)} + 18\mathbf{b}_{21}^{(u)} + 52\mathbf{b}_{00} + 84\mathbf{b}_{01} + 15\mathbf{b}_{02}$$
$$+ 84\mathbf{b}_{10} + 15\mathbf{b}_{20})/412$$
$$\mathbf{c}_{23} = (63\mathbf{b}_{11}^{(u)} + 189\mathbf{b}_{11}^{(v)} + 18\mathbf{b}_{12}^{(u)} + 108\mathbf{b}_{12}^{(v)}$$
$$+ 27\mathbf{b}_{21}^{(u)} + 18\mathbf{b}_{21}^{(v)} + 9\mathbf{b}_{22}^{(u)} + 9\mathbf{b}_{22}^{(v)} + 28\mathbf{b}_{00}$$
$$+ 156\mathbf{b}_{01} + 84\mathbf{b}_{02} + 5\mathbf{b}_{03} + 42\mathbf{b}_{10} + 6\mathbf{b}_{13} + 6\mathbf{b}_{20})/703$$
$$\mathbf{c}_{33} = (234\mathbf{b}_{11}^{(u)} + 234\mathbf{b}_{11}^{(v)} + 108\mathbf{b}_{12}^{(u)} + 144\mathbf{b}_{12}^{(v)} + 144\mathbf{b}_{21}^{(u)} + 108\mathbf{b}_{21}^{(u)}$$
$$+ 63\mathbf{b}_{22}^{(u)} + 14\mathbf{b}_{00} + 84\mathbf{b}_{01} + 42\mathbf{b}_{02} + 2\mathbf{b}_{03} + 84\mathbf{b}_{10} + 15\mathbf{b}_{13}$$
$$+ 42\mathbf{b}_{20} + 6\mathbf{b}_{23} + 2\mathbf{b}_{30} + 15\mathbf{b}_{31} 6\mathbf{b}_{32})/1064$$

and corresponding weights w_{ij}:

$$w_{00} = 0,$$
$$w_{01} = 2/7,$$
$$w_{02} = 11/21,$$
$$w_{03} = 23/35,$$
$$w_{11} = 26/49,$$
$$w_{12} = 109/147$$
$$w_{13} = 211/245$$
$$w_{22} = 412/441$$
$$w_{23} = 256/245$$
$$w_{33} = 282/245.$$

The remaining control points and weights follow by symmetry. The conversion to Bézier form is advantageous for algorithms that benefit from the convex hull property of rational Bézier patches or that need a subdivision process. Note, however, that these patches cannot be formulated in the

IGES specification, since that only allows for strictly positive weights. A way around this is described by Ueda[149].

Gregory patches degenerates to $\mathbf{0}/0$ at the patch corners — this is mirrored by the appearance of zero weights for the corner Bézier points. These singularities are removable: the patch evaluates to the original corner Bézier points.

Parameter values that evaluate to $\mathbf{0}/0$ are called *base points*. These singularities are not always removable: in our case, we have $\mathbf{c}_{01} = \mathbf{c}_{10}$, which is also the patch's corner position. If $\mathbf{c}_{01} \neq \mathbf{c}_{10}$, $(u, v) = (0, 0)$ would be mapped to all values on the straight line between $\mathbf{c}_{01}$ and $\mathbf{c}_{10}$ — a nonremovable singularity. Which value on the line is attained depends on the direction from which $(0, 0)$ is approached as a limit. More on base points in Section 14.5.

14.3   RBG patches

Gregory patches replace the "twist" control points of a bicubic patch by variable points, thus introducing rational terms. One can go a step further and repeat this approach for rational Bézier patches. Then *every* control point and its corresponding weight would be made variable. Such an approach seems overly liberal for the boundary curve control points — it seems reasonable to have those determined in a non-variable fashion. Still, each control point might have two weights. This was carried out by H. Chiyokura et al. in [34]. The corresponding method is called *rational boundary Gregory patch*.

For an explanation, let us group the sixteen control points of a rational bicubic into for blocks, one per patch corner. For the block pertaining to $\mathbf{b}_{00}$, we have the following:

$$\mathbf{b}_{i,j} = \frac{u^2 \mathbf{b}_{i,j}^{(u)} + v^2 \mathbf{b}_{i,j}^{(v)}}{u^2 + v^2}; \quad i, j \in \{0, 1\} \tag{14.2}$$

and

$$w_{i,j} = \frac{u^2 w_{i,j}^{(u)} + v^2 w_{i,j}^{(v)}}{u^2 + v^2}; \quad i, j \in \{0, 1\} \tag{14.3}$$

and analogous requirements for the remaining three blocks. Equations (14.2) actually contain only one "real" equation, namely for $(i, j) = (1, 1)$, since we required all boundary control points to be constant.

An additional requirement is that the weights $w_{i,j}^{(u)}$ are equal for the first and last two rows of control points:

$$w_{i,0}^{(u)} = w_{i,1}^{(u)} \quad \text{and} \quad w_{i,2}^{(u)} = w_{i,3}^{(u)}$$

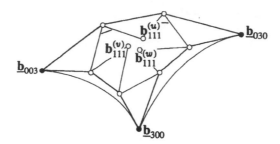

Figure 14.3. Gregory triangles: standard cubic patches have compatibility problems.

for $i = 0, 1, 2, 3$. Similarly, it is required that

$$w_{0,j}^{(v)} = w_{1,j}^{(u)} \quad \text{and} \quad w_{2,j}^{(v)} = w_{3,j}^{(u)}.$$

These conditions allow for easy formulation of C^1 conditions between neighboring patches: the cross-boundary derivative along the edge $v = 0$ is given by

$$\mathbf{x}_v(u, 0) = \frac{3 \sum_{i=0}^{3} B_i^3(u) w_{i,0}^{(u)} [\mathbf{b}_{i,1}^{(u)} - \mathbf{b}_{i,0}^{(u)}]}{\sum_{i=0}^{3} B_i^3(u) w_{i,0}^{(u)}}. \tag{14.4}$$

The RBG concept has been successfully implemented into the DESIGN-BASE modeler, see Chiyokura [32].

14.4   Gregory triangles

Before he invented the rectangular patch that now bears his name, J. Gregory investigated triangular patches.[3] We will present a Bézierbased version of that patch scheme, as described in [104] and [74].

Consider three triangular cubic patches as in Fig. 14.3. If patches A and C are given, can we determine the central control point $\mathbf{b}_{111}$ of patch B so as to ensure C^1 continuity of the overall surface? This is not possible; $\mathbf{b}_{111}$ would be overdetermined.

The Foley/Opitz solution: replace $\mathbf{b}_{111}$ by three independent points $\mathbf{b}_{111}^{(u)}, \mathbf{b}_{111}^{(v)}, \mathbf{b}_{111}^{(w)}$! In the equation for the patch, we will then replace the

---

[3]Private communication. It is also interesting to note that de Casteljau considered triangular patches before he turned to rectangular ones!

constant point $\mathbf{b}_{111}$ by the function

$$\mathbf{b}_{111}(u,v,w) = \frac{vw\mathbf{b}_{111}^{(u)} + uw\mathbf{b}_{111}^{(v)} + uv\mathbf{b}_{111}^{(w)}}{vw + uw + uv}. \qquad (14.5)$$

Again, cross-boundary derivatives across the edge $u = 0$ will only involve one of the three center points, namely $\mathbf{b}_{111}^{(u)}$, with similar statements for the remaining two edges.

It is possible to convert a Gregory triangle into standard rational Bézier form — it will then be a rational triangular patch of degree five. Its Bézier points $\mathbf{c}_{ijk}$ are

$$\begin{aligned}
\mathbf{c}_{050} &= \mathbf{b}_{030}, \\
\mathbf{c}_{041} &= \mathbf{b}_{030}, \\
\mathbf{c}_{032} &= \mathbf{b}_{021}, \\
\mathbf{c}_{131} &= (\mathbf{b}_{030} + 3\mathbf{b}_{021} + 3\mathbf{b}_{120})/7, \\
\mathbf{c}_{122} &= (\mathbf{b}_{021} + \mathbf{b}_{012} + 2\mathbf{b}_{111}^{(u)})/4.
\end{aligned}$$

The corresponding weights w_{ijk} are:

$$\begin{aligned}
w_{050} &= 0, \\
w_{041} &= 1/5, \\
w_{032} &= 3/10, \\
w_{131} &= 7/20, \\
w_{122} &= 2/5.
\end{aligned}$$

All other control points and weights follow by symmetry.

Again, we note base points at the patch corners! The patch has a removable singularity there since the control points $\mathbf{c}_{041}$ and $\mathbf{c}_{140}$ coincide, as do the corresponding ones at the other corners.

The approach by Longhi [104] uses quartic triangles and replaces each of the three interior control points by a weighted combination of two points, thus yielding a total of eighteen control points. We have that

$$\mathbf{b}_{211}(u,v,w) = \frac{(1-v)w\mathbf{b}_{211}^{(w)} + v(1-w)\mathbf{b}_{211}^{(v)}}{(1-v)w + v(1-w)}$$

and analogous expressions for $\mathbf{b}_{121}$ and $\mathbf{b}_{112}$.

14.5   Base points and multi-sided patches

When transcribing a Gregory patch into rational Bézier form, we generated a triangular patch of the following structure (we only describe the corner

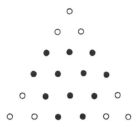

Figure 14.4. Multi-sided patches: by setting some weights zero (open circles), the remaining control points (solid circles) form a six-sided patch. Three of its boundary curves are of degree two, and three are of degree one.

c_{0n0}, the other ones being similar): the corner weight w_{0n0} is zero, and the next two control points $c_{1,n-1,0}$ and $c_{0,n-1,1}$ coincide. If they did not, then our patch would contain the straight line $\overline{c_{1,n-1,0}, c_{0,n-1,1}}$ and similar straight lines at the other corners. Then we would have a six-sided patch! Three of its boundary "curves" would just be straight lines, of course, while the remaining three would be of degree $n - 2$.

We may add more layers of zero weights: if we also set $w_{1,n-1,0} = w_{0,n-1,1} = 0$, then the next layer of control points: $c_{2,n-2,0}, c_{1,n-2,1}, c_{0,n-1,2}$ and their weights would generate a rational quadratic boundary curve of the six-sided patch. This approach of creating multi-sided patches using base points is due to J. Warren [158]. Fig. 14.4 illustrates. These multi-sided patches can be very useful for purposes of filleting and blending.

14.6   Problems

1. Plot the function $z = u/(u + v)$. Study how its behavior near $(0, 0)$ explains the properties of Gregory patches.

2. Then repeat the same for the function $z = vw/(vw + uw + uv)$, but now with u, v, w being barycentric coordinates of a triangle in the plane $z = 0$.

3. Suppose the boundary lines of a bilinear patch do not meet in the patch corners, but miss each other. Devise a bilinear Gregory patch, i.e., one that interpolates to eight "corner points".

15

Examples and standards

This chapter contains practical information for NURBS implementers. It contains the most important data specifications, namely two IGES formats, and several examples of commonly encountered NURBS curves and surfaces. Several illustrations of rational biquadratic representations of quadrics are given in Figures 15.1, 15.2, and 15.3.

15.1 IGES

One of the big obstacles in the CAD/CAM world has been the diversity of data format used by different systems. Ford used Coons patches, General Motors used Gordon surfaces, and Mercedes used bicubic splines, to name just a few. In recent years, there has been a trend toward unification — many systems now produce geometry that adheres to one common data standard: IGES (initial graphics exchange specification). IGES defines most conceivable geometric objects, from points, lines and planes to transformation matrices; here we are only interested in their NURBS formats. More advanced data-exchange formats are presently under way, the most important being STEP, which aims beyond pure geometry: a product is not simply defined by its shape, but also by variables such as material, surface finish, or functionality. STEP integrates such descriptors. Further down the road we should expect specifications that involve variable dimensions, such as the radii of cylinders in a complex assembly.

Figure 15.1. Quadrics: part of a cone represented as four rational patches.

Figure 15.2. Quadrics: part of a sphere represented as four rational patches.

Figure 15.3. Quadrics: a whole sphere represented as eight rational patches.

name	type	description
N	int	number of control points (counted from 0)
n	int	degree
flag1	int	0:nonplanar, 1: planar
flag2	int	0:open, 1: closed
flag3	int	0:rational, 1: integral
flag4	int	0:nonperiodic, 1: periodic
knot sequence		
u_{-n}	double	first knot
$\vdots$	$\vdots$	$\vdots$
u_{L+n}	double	last knot
weights		
w_0	double	first weight
$\vdots$	$\vdots$	$\vdots$
w_N	double	last weight
control points		
$\mathbf{d}_0^x$	double	x−component of first control point
$\mathbf{d}_0^y$	double	y−component of first control point
$\mathbf{d}_0^z$	double	y−component of first control point
$\vdots$	$\vdots$	$\vdots$
$\mathbf{d}_N^x$	double	x−component of last control point
$\mathbf{d}_N^y$	double	y−component of last control point
$\mathbf{d}_N^z$	double	z−component of last control point
range		
u_{start}	double	starting parameter value
u_{end}	double	ending parameter value
plane normal (if flag1=1)		
$\mathbf{n}^x$	double	x−component of normal
$\mathbf{n}^y$	double	y−component of normal
$\mathbf{n}^z$	double	y−component of normal

Table 15.1. IGES format 126: NURBS curves

The IGES NURBS curve format #126 invites several comments:

- All of the knots may occur with multiplicity greater than one. If an interior knot is listed k times, the curve is only guaranteed to be $n - k$ times differentiable there.

name	type	description
M	int	number of control points (counted from 0)in u-direction
N	int	number of control points (counted from 0)in v-direction
m	int	degree in u
n	int	degree in v
flag1	int	1: closed in u–direction, 0: else
flag2	int	1: closed in v–direction, 0: else
flag3	int	0:rational, 1: integral
flag4	int	0:nonperiodic in u, 1: periodic
flag5	int	0:nonperiodic in v, 1: periodic
knot sequences		
u_{-m}	double	first knot in u
$\vdots$	$\vdots$	$\vdots$
u_{M+m}	double	last knot in u
v_{-n}	double	first knot in v
$\vdots$	$\vdots$	$\vdots$
v_{N+n}	double	last knot in v
weights		
$w_{0,0}$	double	first weight
$w_{1,0}$	double	second weight
$\vdots$	$\vdots$	$\vdots$
$w_{M,N}$	double	last weight
control points		
$\mathbf{d}^x_{0,0}$	double	x–component of first control point
$\mathbf{d}^y_{0,0}$	double	y–component of first control point
$\mathbf{d}^z_{0,0}$	double	z–component of first control point
$\mathbf{d}^x_{1,0}$	double	x–component of second control point
$\mathbf{d}^y_{1,0}$	double	y–component of second control point
$\mathbf{d}^z_{1,0}$	double	z–component of second control point
$\vdots$	$\vdots$	$\vdots$
$\mathbf{d}^x_{M,N}$	double	x–component of last control point
$\mathbf{d}^y_{M,N}$	double	y–component of last control point
$\mathbf{d}^z_{M,N}$	double	z–component of last control point
ranges		
u_{start}	double	starting parameter value in u
u_{end}	double	ending parameter value in u
v_{start}	double	starting parameter value in v
v_{end}	double	ending parameter value in v

Table 15.2. IGES format 128: NURBS surfaces

- If the end knots u_n and u_{L+n} are of multiplicity n, the first and last control points will lie on the curve.

- IGES requires one "dummy" knot at the beginning and end of the knot sequence. These have no influence on any computation — they are an IGES design flaw. For example, the knot sequence 0, 0, 0, 0, 1, 1, 1, 1 would be needed to define a cubic Bézier curve. Yet the knot sequence -10, 0, 0, 0, 1, 1, 1, 10 would yield the same curve!

- For periodic curves, IGES offers too many knots, as the beginning and end of the curve is no more special than any point in the interior.

- IGES only allows strictly positive weights; this will cause problems with control vectors (Chapter 5) as well as with periodic curves with simple knots.

- IGES stores 3D points and associated weights; this is *not* the homogeneous form! In order to obtain it, we would have to multiply every point by its corresponding weight.

- IGES also provides a "Directory Entry" in which special properties may be stored, such as being part of a circle, a hyperbola, etc.

Some comments on the IGES 128 surface format(see Table 15.2):

- Since only strictly positive weights are allowed, control vectors (Chapter 5) or Gregory patches (Chapter 14) will cause problems.

- There is a "Directory Entry" to record special features such as being a right cylinder, a torus, or a surface of revolution.

- For issues concerning knot multiplicities, see the comments for the curve format.

- Triangular patches (Chapter 12) are not supported.

15.2   A semicircle

A semicircle may be written as a C^2 piecewise cubic NURB curve with two segments. Assuming a uniform knot sequence, its homogeneous control points are given by:

$$\begin{bmatrix} -1 \\ 0 \\ 1 \end{bmatrix}, \frac{2}{3}\begin{bmatrix} -1 \\ \frac{1}{2} \\ 1 \end{bmatrix}, \frac{1}{3}\begin{bmatrix} 0 \\ 2 \\ 1 \end{bmatrix}, \frac{2}{3}\begin{bmatrix} 1 \\ \frac{1}{2} \\ 1 \end{bmatrix}, \begin{bmatrix} 1 \\ 0 \\ 1 \end{bmatrix}.$$

In the IGES format, we would have to enter the knots as: 0, 0, 0, 1, 2, 2, 2.

15.3   A hyperbolic paraboloid

A triangular patch on a hyperbolic paraboloid is given by

$$\begin{bmatrix} 0 \\ 1 \\ 0 \\ 1 \end{bmatrix}$$

$$\begin{bmatrix} 0 \\ \frac{1}{2} \\ 0 \\ 1 \end{bmatrix} \quad \begin{bmatrix} \frac{1}{2} \\ \frac{1}{2} \\ \frac{1}{2} \\ 1 \end{bmatrix}$$

$$\begin{bmatrix} 0 \\ 0 \\ 0 \\ 1 \end{bmatrix} \quad \begin{bmatrix} \frac{1}{2} \\ 0 \\ 0 \\ 1 \end{bmatrix} \quad \begin{bmatrix} 1 \\ 0 \\ 0 \\ 1 \end{bmatrix}$$

This patch is purely polynomial. Of course, it may also be represented as a *bilinear* patch with control points

$$\begin{bmatrix} 0 \\ 1 \\ 0 \\ 1 \end{bmatrix} \begin{bmatrix} 1 \\ 1 \\ 1 \\ 1 \end{bmatrix} \begin{bmatrix} 0 \\ 0 \\ 0 \\ 1 \end{bmatrix} \begin{bmatrix} 1 \\ 0 \\ 0 \\ 1 \end{bmatrix}.$$

Note that the two diagonals $\mathbf{b}(u, u)$ and $\mathbf{b}(u, 1-u)$ are parabolas, touching the plane at infinity.

15.4   A quarter of a cylinder

A quarter of a cylinder, with the $z-$axis as its axis, is given by the homogeneous biquadratic control net

$$\begin{bmatrix} 1 \\ 0 \\ 0 \\ 1 \end{bmatrix} \sqrt{2}/2 \begin{bmatrix} 1 \\ 1 \\ 0 \\ 1 \end{bmatrix} \begin{bmatrix} 1 \\ 0 \\ 0 \\ 1 \end{bmatrix}$$

$$\begin{bmatrix} 1 \\ 0 \\ 1 \\ 1 \end{bmatrix} \sqrt{2}/2 \begin{bmatrix} 1 \\ 1 \\ 1 \\ 1 \end{bmatrix} \begin{bmatrix} 1 \\ 0 \\ 1 \\ 1 \end{bmatrix}$$

If we let all points with $z-$component equal to 1 collapse into one point $[0, 0, 1, 1]^T$, we obtain a quarter of a cone.

15.5 A rational quadratic patch on a sphere

The following six homogeneous control points and weights (quoted from Niebuhr [110]) describe a rational quadratic triangular patch on a sphere:

$$\begin{bmatrix} 0 \\ 0 \\ 1 \\ 1 \end{bmatrix}$$

$$\begin{bmatrix} \frac{1}{2}\sqrt{2} \\ 0 \\ \frac{1}{2}\sqrt{2} \\ \frac{1}{2}\sqrt{2} \end{bmatrix} \begin{bmatrix} \frac{1}{2} \\ \frac{1}{2} \\ \frac{1}{2} \\ \frac{1}{2} \end{bmatrix}$$

$$\begin{bmatrix} 1 \\ 0 \\ 0 \\ 1 \end{bmatrix} \begin{bmatrix} \frac{1}{2}\sqrt{2} \\ \frac{1}{2}\sqrt{2} \\ 0 \\ \frac{1}{2}\sqrt{2} \end{bmatrix} \begin{bmatrix} 0 \\ 1 \\ 0 \\ 1 \end{bmatrix}$$

Such patches are obtainable from a stereographic projection; in this case, its center $\underline{c}$ is given by

$$\underline{c} = \begin{bmatrix} -1 \\ 0 \\ 0 \\ 1 \end{bmatrix}.$$

15.6 An octant of a sphere

We give a rational quartic representation of an octant of the sphere as a triangular patch (Farin, Piper, Worsey [60]). The control net (in homogeneous form) is given by

$$w_{040} \begin{bmatrix} 0 \\ 0 \\ 1 \\ 1 \end{bmatrix}$$

$$w_{031} \begin{bmatrix} \alpha \\ 0 \\ 1 \\ 1 \end{bmatrix} \quad w_{130} \begin{bmatrix} 0 \\ \alpha \\ 1 \\ 1 \end{bmatrix}$$

$$w_{022} \begin{bmatrix} \beta \\ 0 \\ \beta \\ 1 \end{bmatrix} \quad w_{121} \begin{bmatrix} \gamma \\ \gamma \\ 1 \\ 1 \end{bmatrix} \quad w_{220} \begin{bmatrix} 0 \\ \beta \\ \beta \\ 1 \end{bmatrix}$$

$$w_{013} \begin{bmatrix} 1 \\ 0 \\ \alpha \\ 1 \end{bmatrix} \quad w_{112} \begin{bmatrix} 1 \\ \gamma \\ \gamma \\ 1 \end{bmatrix} \quad w_{211} \begin{bmatrix} \gamma \\ 1 \\ \gamma \\ 1 \end{bmatrix} \quad w_{310} \begin{bmatrix} 0 \\ 1 \\ \alpha \\ 1 \end{bmatrix}$$

$$w_{004} \begin{bmatrix} 1 \\ 0 \\ 0 \\ 1 \end{bmatrix} \quad w_{103} \begin{bmatrix} 1 \\ \alpha \\ 0 \\ 1 \end{bmatrix} \quad w_{202} \begin{bmatrix} \beta \\ \beta \\ 0 \\ 1 \end{bmatrix} \quad w_{301} \begin{bmatrix} \alpha \\ 1 \\ 0 \\ 1 \end{bmatrix} \quad w_{400} \begin{bmatrix} 0 \\ 1 \\ 0 \\ 1 \end{bmatrix}$$

where

$$\alpha = (\sqrt{3} - 1)/\sqrt{3}, \quad \beta = (\sqrt{3} + 1)/2\sqrt{3}, \quad \gamma = 1 - (5 - \sqrt{2})(7 - \sqrt{3})/46.$$

The weights are:

$$w_{040} = 4\sqrt{3}(\sqrt{3} - 1), w_{031} = 3\sqrt{2}, w_{202} = 4, w_{121} = \frac{\sqrt{2}}{\sqrt{3}}(3 + 2\sqrt{2} - \sqrt{3}),$$

the other ones following by symmetry.

In order to represent the whole sphere, we would assemble eight copies of this octant patch. Other representations are also possible: each octant may be written as a rational biquadratic patch (introducing singularities at the north and south poles), see [121].

15.7   A sixth of a sphere

The following is due to J. Cobb [36]. The sphere is covered by six rectangular rational biquartic patches, creating a cube-like topology. Each of those patches may be obtained by appropriately rotating the given control net. We give the homogeneous form.

$$
\begin{bmatrix} 4(1-t) \\ 4(1-t) \\ 4(1-t) \\ 4(3-t) \end{bmatrix}
\begin{bmatrix} -d \\ d(t-4) \\ d(t-4) \\ d(3t-2) \end{bmatrix}
\begin{bmatrix} 0 \\ 4(1-2t)/3 \\ 4(1-2t)/3 \\ 4(5-t)/3 \end{bmatrix}
\begin{bmatrix} d \\ d(t-4) \\ d(t-4) \\ d(3t-2) \end{bmatrix}
\begin{bmatrix} 4(t-1) \\ 4(1-t) \\ 4(1-t) \\ 4(3-t) \end{bmatrix}
$$

$$
\begin{bmatrix} d(t-4) \\ -d \\ d(t-4) \\ d(3t-2) \end{bmatrix}
\begin{bmatrix} (2-3t)/2 \\ (2-3t)/2 \\ -(t+6)/2 \\ (t+6)/2 \end{bmatrix}
\begin{bmatrix} 0 \\ d(2t-7)/3 \\ -5\sqrt{6}/3 \\ d((t+6)/3 \end{bmatrix}
\begin{bmatrix} (3t-2)/2 \\ (2-3t)/2 \\ -(t+6)/2 \\ (t+6)/2 \end{bmatrix}
\begin{bmatrix} d(4-t) \\ -d \\ d(t-4) \\ d(3t-2) \end{bmatrix}
$$

$$
\begin{bmatrix} 4(1-2t)/3 \\ 0 \\ 4(1-2t)/3 \\ 4(5-t)/3 \end{bmatrix}
\begin{bmatrix} d(2t-7)/3 \\ 0 \\ -5\sqrt{6}/3 \\ d(t+6)/3 \end{bmatrix}
\begin{bmatrix} 0 \\ 0 \\ 4(t-5)/3 \\ 4(5t-1)/9 \end{bmatrix}
\begin{bmatrix} -d(2t-7)/3 \\ 0 \\ -5\sqrt{6}/3 \\ d(t+6)/3 \end{bmatrix}
\begin{bmatrix} -4(1-2t)/3 \\ 0 \\ 4(1-2t)/3 \\ 4(5-t)/3 \end{bmatrix}
$$

$$
\begin{bmatrix} d(t-4) \\ -d \\ d(t-4) \\ d(3t-2) \end{bmatrix}
\begin{bmatrix} (2-3t)/2 \\ -(2-3t)/2 \\ -(t+6)/2 \\ (t+6)/2 \end{bmatrix}
\begin{bmatrix} 0 \\ -d(2t-7)/3 \\ -5\sqrt{6}/3 \\ d((t+6)/3 \end{bmatrix}
\begin{bmatrix} (3t-2)/2 \\ -(2-3t)/2 \\ -(t+6)/2 \\ (t+6)/2 \end{bmatrix}
\begin{bmatrix} d(4-t) \\ d \\ d(t-4) \\ d(3t-2) \end{bmatrix}
$$

$$
\begin{bmatrix} 4(1-t) \\ -4(1-t) \\ 4(1-t) \\ 4(3-t) \end{bmatrix}
\begin{bmatrix} -d \\ -d(t-4) \\ d(t-4) \\ d(3t-2) \end{bmatrix}
\begin{bmatrix} 0 \\ -4(1-2t)/3 \\ 4(1-2t)/3 \\ 4(5-t)/3 \end{bmatrix}
\begin{bmatrix} d \\ -d(t-4) \\ d(t-4) \\ d(3t-2) \end{bmatrix}
\begin{bmatrix} 4(t-1) \\ -4(1-t) \\ 4(1-t) \\ 4(3-t) \end{bmatrix}
$$

where $d = \sqrt{2}$ and $t = \sqrt{3}$.

15.8 A sixteenth of a torus

A torus is a surface of revolution. We consider a torus that has been created by revolving a circle around the z–axis. Half of the torus is above the $z = 0$ plane, the other half is below it. Here are the homogeneous control points for an "interior" patch with $x, y, z \geq 0$, writing it in homogeneous biquadratic form:

$$
\begin{bmatrix} 1 \\ 0 \\ 0 \\ 1 \end{bmatrix}
\begin{bmatrix} \frac{1}{2}\sqrt{2} \\ \frac{1}{2}\sqrt{2} \\ 0 \\ \frac{1}{2}\sqrt{2} \end{bmatrix}
\begin{bmatrix} 0 \\ 1 \\ 0 \\ 1 \end{bmatrix}
\begin{bmatrix} \frac{1}{2}\sqrt{2} \\ 0 \\ \frac{1}{2}\sqrt{2} \\ \frac{1}{2}\sqrt{2} \end{bmatrix}
\begin{bmatrix} \frac{1}{2} \\ \frac{1}{2} \\ \frac{1}{2} \\ \frac{1}{2} \end{bmatrix}
\begin{bmatrix} 0 \\ \frac{1}{2}\sqrt{2} \\ \frac{1}{2}\sqrt{2} \\ \frac{1}{2}\sqrt{2} \end{bmatrix}
$$

$$
\begin{bmatrix} 2 \\ 0 \\ 0 \\ 1 \end{bmatrix}
\begin{bmatrix} \sqrt{2} \\ \sqrt{2} \\ \frac{1}{2}\sqrt{2} \\ \frac{1}{2}\sqrt{2} \end{bmatrix}
\begin{bmatrix} 0 \\ 2 \\ 1 \\ 1 \end{bmatrix}
$$

Just for the record: a torus may have four intersections with a straight line — hence it is *not* a quadric!

15.9   Problems

1. The hyperbolic paraboloid from Section 15.3 is obtainable from a stereographic projection. What is its center?

2. Redo Section 15.3 and the previous problem for an *elliptic paraboloid*.

3. Write a semisphere as a rational triangular cubic patch.

16

Notation

Here is the notation used in this book:

$\wedge$	cross product; $\underline{\mathbf{L}} \wedge \underline{\mathbf{M}}$: intersection of two lines; $\underline{\mathbf{p}} \wedge \underline{\mathbf{q}}$: line through two points
$\wedge$	cross product
$\dot{}, \ddot{}$	curve derivatives with respect to the current parameter
a, b, α, β	real numbers or real-valued functions
$\mathbf{0}$	short for (0,0,0)
$\mathbf{a}, \mathbf{b}$	points or vectors in affine space
$\underline{\mathbf{a}}, \underline{\mathbf{b}}$	points in projective space
	If $\underline{\mathbf{b}} = [x, y, z, w]^{\mathrm{T}}$, then $\mathbf{b} = [x/w, y/w, z/w]^{\mathrm{T}}$.
A, B	matrices
$\mathbf{L}, \mathbf{M}$	lines in projective space
$B_i^n, B_{\mathbf{i}}^n$	Bernstein polynomials of degree n
$\mathbf{e1}, \mathbf{e2}, \mathbf{e3}$	short for $[1, 0, 0]^{\mathrm{T}}, [0, 1, 0]^{\mathrm{T}}, and [0, 0, 1]^{\mathrm{T}}$
$I\!E^d$	$d-$dimensional euclidean space
$D_{\mathbf{d}}f$	directional derivative of f in the direction $\mathbf{d}$
Δ_i	difference in parameter intervals (i.e., $\Delta_i = u_{i+1} - u_i$)
Δ^r	iterated forward difference
H_i^3	cubic Hermite polynomials
$\mathbf{P}$	control polygon

Φ	an affine or projective map
λ	an interval in a knot sequence
$\|\mathbf{v}\|$	(euclidean) length of the vector $\mathbf{v}$
$\mathbf{x}_u$	u-partial of $\mathbf{x}(u, v)$

Bibliography

[1] S. Abhyankar and C. Bajaj. Automatic parametrization of rational curves and surfaces I: conics and conicoids. *Computer Aided Design*, 19(1), 1987.

[2] S. Abhyankar and C. Bajaj. Automatic parametrization of rational curves and surfaces II: cubics and cubicoids. *Computer Aided Design*, 19(9), 1987.

[3] S. Abhyankar and C. Bajaj. Automatic parametrization of rational curves and surfaces III: Algebraic plane curves. *Computer Aided Geometric Design*, 5(4):309–322, 1988.

[4] S. Abhyankar and C. Bajaj. Automatic parametrization of rational curves and surfaces IV. *ACM Transactions on Graphics*, 8(4):325–334, 1989.

[5] D. Ahuja and S. Coons. Geometry for construction and display. *IBM System Journal*, 7(3/4):188–205, 1968.

[6] G. Albrecht. A remark on Farin points for Bézier triangles. *cagd*, 12(5):507–512, 1995.

[7] G. Albrecht. Determination and classification of triangular quadric patches. *Computer Aided Geometric Design*, 15(7):675–69, 1998.

[8] G. Aumann. Interpolation with developable Bezier patches. *Computer Aided Geometric Design*, 8(5):409–420, 1991.

[9] H. Baker. *Principles of Geometry, Vol. III*. Cambridge Univ. Press, London, 1934.

[10] A. Ball. Consurf I: introduction of the conic lofting tile. *Computer Aided Design*, 6(4):243–249, 1974.

[11] A. Ball. Consurf II: description of the algorithms. *Computer Aided Design*, 7(4):237–242, 1975.

[12] A. Ball. Consurf III: how the program is used. *Computer Aided Design*, 9(1):9–12, 1977.

[13] L. Bardis and N. Patrikalakis. Approximate conversion of rational B-spline patches. *Computer Aided Geometric Design*, 6(3):189–204, 1989.

[14] R. Barnhill. Coons' patches. *Computers in Industry*, 3:37–43, 1982.

[15] R. Barnhill and J. Gregory. Compatible smooth interpolation in triangles. *J of Approx. Theory*, 15(3):214–225, 1975.

[16] R. Bartels, J. Beatty, and B. Barsky. *An Introduction to Splines for Use in Computer Graphics and Geometric Modeling*. Morgan Kaufmann, 1987.

[17] C. Baumgarten and G. Farin. Rational approximation of spirals. *Computer Aided Geometric Design*, 14(6):515–532, 1997.

[18] M. Berger. *Geometry I*. Springer-Verlag, 1987.

[19] D. Binschadler and W. Frey. Developable surfaces for math-based binder design, 1993. GMR report MA-526.

[20] J. Blinn and M. Newell. Clipping using homogeneous coordinates. *Computer Graphics*, 12(3):281–287, 1978.

[21] R. Bodduluri and B. Ravani. Design of developable surfaces using duality between plane and point geometries. *Computer Aided Design*, 25(10):621–632, 1993.

[22] W. Boehm. On cubics: a survey. *Computer Graphics and Image Processing*, 19:201–226, 1982.

[23] W. Boehm. An affine represenation of de casteljau's and de boor's rational algorithms. *Computer Aided Geometric Design*, 10(3-4):175–180, 1993.

[24] W. Boehm. Some remarks on quadrics. *Computer Aided Geometric Design*, 10(3-4):231–236, 1993.

[25] W. Boehm and D. Hansford. Bézier patches on quadrics. In G. Farin, editor, *NURBS for Curve and Surface Design*, pages 1–14. SIAM, 1991.

[26] W. Boehm and H. Prautzsch. *Geometric Foundations of Geometric Design*. AK Peters, Boston, 1992.

[27] G. Bol. *Projective Differential Geometry, Vol. 1*. Vandenhoeck and Ruprecht, Goettingen, 1950. Vol. 2in 1954, Vol.3 in 1967. In German.

[28] K. Bolton. Biarc curves. *Computer Aided Design*, 7(2):89–92, 1975.

[29] G. Bonneau. Weight estimation of rational Bezier curves and surfaces. In H. Hagen, G. Farin, and H. Noltemeier, editors, *Geometric Modeling*, pages 79–86. Springer, Vienna, 1995.

[30] G. Brunnett, T. Schreiber, and J. Braun. The geometry of optimal degree reduction of Bézier curves. *Computer Aided Geometric Design*, 13(8):773–788, 1996.

[31] F. Chen. Bilinear precision of rational Bézier surfaces. *Computer Aided Geometric Design*, 14(4):387–392, 1997.

[32] H. Chiyokura. *Solid Modeling with DESIGNBASE*. Addison-Wesley, 1988.

[33] H. Chiyokura and F. Kimura. Design of solids with free-form surfaces. *Computer Graphics*, 17(3):289–298, 1983.

[34] H. Chiyokura, T. Takamura, K. Konno, and T. Harada. g^1 surface interpolation over irregular meshes with rational curves. In G. Farin, editor, *NURBS for Curve and Surface Design*, pages 15–34. SIAM, 1991.

[35] J. Chou. Higher order bézier circles. *Computer Aided Design*, 27(4):303–309, 1995.

[36] J. Cobb. A rational bicubic representation of the sphere. Technical report, Computer science, U. of Utah, 1988.

[37] H. Coxeter. *The Real Projective Plane*. Cambridge Univ. Press, 1961.

[38] P. Davis. *Interpolation and Approximation*. Dover, New York, 1975. first edition 1963.

[39] C. de Boor. On calculating with B-splines. *J Approx. Theory*, 6(1):50–62, 1972.

[40] C. de Boor. *A Practical Guide to Splines*. Springer, 1978.

[41] C. de Boor, K. Höllig, and M. Sabin. High accuracy geometric Hermite interpolation. *Computer Aided Geometric Design*, 4(4):269–278, 1987.

[42] P. de Casteljau. Outillages méthodes calcul. Technical report, A. Citroen, Paris, 1959.

[43] P. de Casteljau. Courbes et surfaces à poles. Technical report, A. Citroen, Paris, 1963.

[44] W. L. F. Degen. The types of triangular bézier surfaces. In Glen Mullineux, editor, *The Mathematics of Surfaces VI*, pages 153–170. Oxford University Press, 1996.

[45] R. Dietz, J. Hoschek, and B. Juettler. An algebraic approach to curves and surfaces on the sphere and on other quadrics. *Computer Aided Geometric Design*, 10(3-4):211–229, 1993.

[46] M. do Carmo. *Differential Geometry of Curves and Surfaces*. Prentice Hall, Englewood Cliffs, 1976.

[47] M. Eck. Degree reduction of Bézier curves. *Computer Aided Geometric Design*, 10(3-4):237–252, 1993.

[48] M. Eck. Least squares degree reduction of Bézier curves. *Computer Aided Design*, 27(11):845–851, 1995.

[49] G. Farin. Konstruktion und Eigenschaften von Bézier-Kurven und -Flächen. Master's thesis, Technical University Braunschweig, FRG, 1977.

[50] G. Farin. Algorithms for rational Bézier curves. *Computer Aided Design*, 15(2):73–77, 1983.

[51] G. Farin. Triangular Bernstein-Bézier patches. *Computer Aided Geometric Design*, 3(2):83–128, 1986.

[52] G. Farin. Curvature continuity and offsets for piecewise conics. *ACM Transactions on Graphics*, 8(2):89–99, 1989.

[53] G. Farin. Rational curves and surfaces. In T. Lyche and L. Schumaker, editors, *Mathematical Aspects in CAGD*, pages 215–238. Academic Press, 1989.

[54] G. Farin. Tighter convex hulls for rational Bézier curves. *Computer Aided Geometric Design*, 10(2):123–126, 1993.

[55] G. Farin. Projective blossoms and derivatives. In *Curves and Surfaces in Geometric Design*, pages 147–152. AK Peters, Boston, 1994.

[56] G. Farin. The geometry of C^1 projective curves and surfaces. In *Mathematical Methods for Curves and Surfaces*, pages 133–139. Vanderbilt Univ. Press, Nashville, 1995.

[57] G. Farin. *Curves and Surfaces for Computer Aided Geometric Design*. Academic Press, 1996. fourth edition.

[58] G. Farin. Projective splines. In P. Kaklis and N. Sapidis, editors, *Computer Aided Geometric Design*, pages 1–8. National Technical University of Athens, 1996.

[59] G. Farin and D. Jung. Linear precision of rational Bézier curves. *Computer Aided Geometric Design*, 12(4):431–433, 1995.

[60] G. Farin, B. Piper, and A. Worsey. The octant of a sphere as a non-degenerate triangular Bézier patch. *Computer Aided Geometric Design*, 4(4):329–332, 1988.

[61] R. Farouki. Pythagorean-hodograph curves in practical use. In R. E. Barnhill, editor, *Geometry Processing for Design and Manufacturing*, pages 3–33. SIAM, Philadelphia, 1992.

[62] R. Farouki, J. Manjunathaiah, D. Nicholas, G. Yuan, and S. Jee. Variable-feedrate CNC interpolators for constant material removal rates along pythagorean-hodograph curves. *Computer Aided Design*, 30(8):631–640, 1998.

[63] R. Farouki and T. Sakkalis. Pythagorean hodographs. *IBMmJournal of Research and development*, 34:736–752, 1990.

[64] R. Farouki and T. Sakkalis. Real rational curves are not 'unit speed'. *Computer Aided Geometric Design*, 8(2):151–158, 1991.

[65] R. Farouki and S. Shah. Real-time CNC interpolators for Pythagorean-hodograph curves. *Computer Aided Geometric Design*, 13(5):583–600, 1996.

[66] R. Farouki, Y. Tsai, and G. Yuan. Contour machining of free-form surfaces with real-time PH curve CNC interpolators. *Computer Aided Geometric Design*. to appear.

[67] R. T. Farouki and H. Pottmann. Polynomial and rational pythagorian-hodograph curves reconciled. In Glen Mullineux, editor, *The Mathematics of Surfaces VI*, pages 355–378. Oxford University Press, 1996.

[68] J. Fiorot and P. Jeannin. *Rational Curves and Surfaces*. Wiley, Chicester, 1992. Translated from the French by M. Harrison.

[69] J. Fiorot and P. Jeannin. Linear precision of BR-curves. *Computer Aided Geometric Design*, 12(4):435–438, 1995.

[70] J. Fiorot, P. Jeannin, and S. Taleb. Rational cubic curves as BR-curves. *Computer Aided Geometric Design*, 12(1):67–88, 1995.

[71] M. Floater. Evaluation and properties of the derivatives of a NURBS curve. In *Mathematical Methods in CAGD II*, pages 261–274. Academic Press, 1992.

[72] M. Floater. Derivatives of rational Bézier curves. *Computer Aided Geometric Design*, 10, 1993.

[73] J. Foley and A. Van Dam. *Fundamentals of Interactive Computer Graphics*. Addison-Wesley, 1982.

[74] T. Foley and K. Opitz. Hybrid cubic Bézier patches. In *Mathematical Methods in CAGD II*, pages 275–286. Academic Press, Boston, 1992.

[75] A. Forrest. *Curves and surfaces for computer-aided design*. PhD thesis, Cambridge, 1968.

[76] A. Forrest. Interactive interpolation and approximation by Bézier polynomials. *The Computer J*, 15(1):71–79, 1972. reprinted in CAD 22(9):527–537,1990.

[77] A. Forrest. The twisted cubic curve: a computer-aided geometric design approach. *Computer Aided Design*, 12(4):165–172, 1980.

[78] I. Fudos and C. Hoffmann. Constraint-based parametric conics for CAD. *Computer Aided Design*, 28(2), 1996.

[79] G. Geise and B. Juettler. A geometrical approach to curvature continuous joints of rational curves. *Computer Aided Geometric Design*, 10(2):109–122, 1993.

[80] R. Goldman. The method of resolvents: a technique for the implicitization, inversion, and intersection of non-planar, parametric, rational cubic curves. *Computer Aided Geometric Design*, 2(4):237–255, 1985.

[81] R. Goldman, T. Sederberg, and D. Anderson. Vector elimination: A technique for the implicitization, inversion, and intersection of planar parametric rational polynomial curves. *Computer Aided Geometric Design*, 1(4):327–356, 1984.

[82] T. Goodman. Shape preserving interpolation by parametric rational cubic splines. Technical report, University of Dundee, 1988. Department of Mathematics and Computer Science.

[83] J. Gregory. Smooth interpolation without twist constraints. In R. E. Barnhill and R. F. Riesenfeld, editors, *Computer Aided Geometric Design*, pages 71–88. Academic Press, 1974.

[84] D. Hansford. *Boundary curves with quadric precsion for a tangent continuous scattered data interpolant.* PhD thesis, Computer Science, Arizona State Univ., 1991.

[85] T. Hermann. G^2 interpolation of free form curve networks by biquintic Gregory patches. *Computer Aided Geometric Design*, 13(9):873–894, 1996.

[86] D. Hilbert and S. Cohn-Vossen. *Geometry and the Imagination.* Chelsea, New York, 1952.

[87] J. Hoschek. Dual Bézier curves and surfaces. In R. Barnhill and W. Boehm, editors, *Surfaces in Computer Aided Geometric Design*, pages 147–156. North-Holland, 1983.

[88] J. Hoschek. Bézier curves and surfaces on quadrics. In *Curves and Surfaces in CAGD*, pages 1–4. Academic Press, 1992.

[89] J. Hoschek. Circular splines. *Computer Aided Design*, 24(6):611–618, 1992.

[90] J. Hoschek and D. Lasser. *Grundlagen der Geometrischen Datenverarbeitung.* B.G. Teubner, Stuttgart, 1989. English translation: Fundamentals of Computer Aided Geometric Design, AK Peters, 1993.

[91] J. Hoschek and F. Schneider. Spline conversion for trimmed rational Bézier- and B-spline surfaces. *Computer Aided Design*, 22(9):580–590, 1990.

[92] J. Hoschek and F.-J. Schneider. Approximate spline conversion for integral and rational Bézier and B-Spline surfaces. In R. E. Barnhill, editor, *Geometry Processing for Design and Manufacturing*, pages 45–86. SIAM, Philadelphia, 1992.

[93] S. Jolles. *Die Theorie der Oskulanten und das Sehnensystem der Raumkurve 4. Ordnung, 2. Spezies.* PhD thesis, Technical Univ. Aachen, 1886.

[94] T. Kim. *Design of rational curves and developable surfaces.* PhD thesis, Arizona State University, 1996.

[95] R. Klass. An offset spline approximation for plane cubics. *Computer Aided Design*, 15(5):296–299, 1983.

[96] K. Konno and H. Chiyokura. An approach of desgiand and controlling free-form surfaces by using NURBS boundary Gregory patches. *Computer Aided Geometric Design*, 13(9):825–850, 1996.

[97] K. Kubota. Pythagorean triples in unique factorization domains. *American Mathematics Monthly*, 79:503–505, 1972.

[98] M. Lachance. Chebychev economization for parametric surfaces. *Computer Aided Geometric Design*, 5(3):195–208, 1988.

[99] M. Lachance and A. Schwartz. Four point parabolic interpolation. *Computer Aided Geometric Design*, 8(2):143–150, 1991.

[100] E. Lee and M. Lucian. Moebius reparametrizations of rational B-splines. *Computer Aided Geometric Design*, 8(3):213–216, 1991.

[101] S. Leopoldseder and H. Pottmann. Approximation of developable surfaces with cone spline surfaces. *Computer Aided Design*, 30(7):571–582, 1998.

[102] R. Liming. *Practical analytical geometry with applications to aircraft.* Macmillan, 1944.

[103] R. Liming. *Mathematics for Computer Graphics.* Aero publishers, 1979.

[104] L. Longhi. Interpolating patches between cubic boundaries. Technical report, Univ. of California, Berkeley, 1986. Technical Report T.R. UCB/CSD 87/313.

[105] K. Matsumura, A. Yamada, and F. Yamaguchi. A geometric newton method for interference processing of rational curves and surfaces in homogeneous space. In Glen Mullineux, editor, *The Mathematics of Surfaces VI*, pages 379–396. Oxford University Press, 1996.

[106] D. Meek and D. Walton. Approximation of discrete data by G^1 arc splines. *Computer Aided Design*, 24(6):301–306, 1992.

[107] K. Miura and K. Wang. Gregory-type patches bounded by low-degree integral curves for G^2 continuity. *Computer Aided Geometric Design*, 13(9):793–810, 1996.

[108] F. Moebius. *August Ferdinand Moebius, Gesammelte Werke.* Verlag von S. Hirzel, 1885. also published by Dr. M. Saendig oHG, Wiesbaden, FRG, 1967.

[109] L. J. Nachman. Matching nurbs surfaces g^1 and g^2. In Glen Mullineux, editor, *The Mathematics of Surfaces VI*, pages 475–516. Oxford University Press, 1996.

[110] M. Niebuhr. *Properties of quadrics in rational Bézier form.* PhD thesis, TU Braunschweig, Germany, 1992. in German.

[111] C. J. Ong, Y. S. Wong, H. T. Loh, and X. G. Hong. An optimization approach for biarc curve-fitting of B-spline curves. *Computer Aided Design*, 28(12), 1996.

[112] D. Parkinson. Optimal biarc curves with tension. *Computer Aided Geometric Design*, 9(3):207–218, 1992.

[113] D. Parkinson and D. Moreton. Optimal biarc-curve fitting. *Computer Aided Design*, 23(6):411–419, 1991.

[114] N. Patrikalakis. Approximate conversion of rational splines. *Computer Aided Geometric Design*, 6(2):155–165, 1989.

[115] D. Pedoe. *A course of Geometry.* Cambridge Univ. Press, London, 1970.

[116] M. Penna and R. Patterson. *Projective Geometry and its Applications to Computer Graphics.* Prentice Hall, 1986.

[117] J. Peters and U. Reif. The 42 equivalence classes of quadratic surfaces in affine n-space. *Computer Aided Geometric Design*, 15(5):459–474, 1998.

[118] L. Piegl. On the use of infinite control points in CAGD. *Computer Aided Geometric Design*, 4(1-2):155–166, 1987.

[119] L. Piegl. Modifying the shape of rational B-splines Part 1: curves. *Computer Aided Design*, 21(8):509–518, 1989.

[120] L. Piegl. On NURBS: a survey. *Computer Graphics and Applications*, 11(1):55–71, 1990.

[121] L. Piegl and W. Tiller. Curve and surface constructions using rational B-splines. *Computer Aided Design*, 19(9):485–498, 1987.

[122] L. Piegl and W. Tiller. Algorithm for approximate nurbs skinning. *Computer Aided Design*, 28(9), 1996.

[123] H. Pottmann. Locally controllable conic splines with curvature continuity. *ACM Transactions on Graphics*, 10(4):366–377, 1991.

[124] H. Pottmann. Rational curves and surfaces with rational offsets. *Computer Aided Geometric Design*, 12(2):175–192, 1995.

[125] H. Pottmann and G. Farin. Developable rational Bézier and B-spline surfaces. *Computer Aided Geometric Design*, 12(5):513–531, 1995.

[126] H. Qin and D. Terzopoulos. Physics-based NURBS swung surfaces. In Glen Mullineux, editor, *The Mathematics of Surfaces VI*, pages 267–290. Oxford University Press, 1996.

[127] L. Ramshaw. Blossoming: a connect-the-dots approach to splines. Technical report, Digital Systems Research Center, Palo Alto, Ca, 1987.

[128] L. Ramshaw. Blossoms are polar forms. *Computer Aided Geometric Design*, 6(4):323–359, 1989.

[129] P. Redont. Representation and deformation of developable surfaces. *Computer Aided Design*, 21(1):13–20, 1989.

[130] R. Riesenfeld. Homogeneous coordinates and projective planes in computer graphics. *IEEE Computer Graphics and Applications*, 1:50–55, 1981.

[131] M. Rowin. Conic, cubic, and T-conic segments. Technical Report Document D2-23252, The Boeing Co., 1964.

[132] M. Sabin. *The use of piecewise forms for the numerical representation of shape*. PhD thesis, Hungarian Academy of Sciences, Budapest, Hungary, 1976.

[133] T. Saito, G. Wang, and T. Sederberg. Hodographs and normals of rational curves and surfaces. *Computer Aided Geometric Design*, 12(4):417–430, 1995.

[134] T. Sederberg. *Implicit and parametric curves and surfaces for computer aided geometric design*. PhD thesis, Mech. Eng., Purdue U., 1983.

[135] T. Sederberg. Improperly parametrized rational curves. *Computer Aided Geometric Design*, 3(1):67–75, 1986.

[136] T. Sederberg. Point and tangent computation of tensor product rational Bézier surfaces. *Computer Aided Geometric Design*, 12(1):103–106, 1995.

[137] T. Sederberg and D. Anderson. Steiner surface patches. *IEEE Computer Graphics and Applications*, 5(5):23–36, 1985.

[138] T. Sederberg, D. Anderson, and R. Goldman. Implicit representation of parametric curves and surfaces. *Computer Vision, Graphics, and Image Processing*, 28(1):72–84, 1984.

[139] T. Sederberg and M. Kakimoto. Approximating rational curves using polymial curves. In G. Farin, editor, *NURBS for Curve and Surface Design*, pages 149–158. SIAM, 1991.

[140] T. Sederberg and X. Wang. Rational hodographs. *Computer Aided Geometric Design*, 4(4):333–335, 1987.

[141] T. Sharrock. Biarcs in three dimensions. In R. Martin, editor, *The Mathematics of Surfaces II*, pages 395–412. Oxford University Press, 1987.

[142] K. Spitzmueller. Partial derivatives of Bézier surfaces. *Computer Aided Design*, 28(1), 1996.

[143] J. Stolfi. *Oriented Projective Geometry*. Academic Press, 1991.

[144] D. Struik. *Analytic and Projective Geometry*. Addison-Wesley, Cambridge, MA, 1953.

[145] T. Takamura, M. Ohta, H. Toriya, and H. Chiyokura. A method to convert a Gregory patch and rational boundary Gregory patch to a rational Bézier patch and its applications. In *Proceedings of Computer Graphics International '90*, pages 543–562. Springer, 1990.

[146] S. Teller and C. Sequin. Modeling implicit quadrics and free-form surfaces with trimmed rational quadratic Bézier patches. Technical report, Computer Science, Berkeley, 1990. report no. UCB/CSD 90/577.

[147] R. Tookey and A. Ball. Approximate g^1 continuous interpolation of a rectangular network of rational cubic curves. *Computer Aided Design*, 28(12), 1996.

[148] D. M. Tsay and B. J. Lin. Improving the geometry design of cylindrical cams using nonparametric rational B-splines. *Computer Aided Design*, 28(1), 1996.

[149] K. Ueda. A method for removing the singularities from Gregory surfaces. In *Mathematical Methods in CAGD II*. Academic Press, 1992.

[150] K. Ueda. A modification to the rational boundary Gregory patch. *Computer Aided Geometric Design*, 13(9):811–824, 1996.

[151] K. Vesprille. *Computer aided design applications of the rational B-spline approximation form*. PhD thesis, Syracuse U., 1975.

[152] D. Walton and D. Meek. G^2 curves composed of planar cubic an Pythagorean hodograph quintic spirals. *Computer Aided Geometric Design*, 15(6):547–566, 1998.

[153] D. J. Walton and D. S. Meek. A pythagorean hodograph quintic spiral. *Computer Aided Design*, 28(12), 1996.

[154] G. Wang, T. Sederberg, and T. Saito. Partial derivatives of rational Bézier surfaces. *Computer Aided Geometric Design*, 14(4):377–382, 1997.

[155] G.-J. Wang and G.-Z. Wang. The rational cubic Bézier representation of conics. *Computer Aided Geometric Design*, 9(6):447–456, 1992.

[156] L. Wang. Coons type blended B-spline surface and its conversion to NURBS surface. *Computer and Graphics*, 21(3):387–401, 1997.

[157] W. Wang and B. Joe. Interpolation on quadric surfaces with rational quadratic spline curves. *Computer Aided Geometric Design*, 14(3):207–230, 1997.

[158] J. Warren. Creating multisided rational Bézier surfaces using base points. *ACM Transactions on Graphics*, 11(2):127–139, 1992.

[159] M. Watkins and A. Worsey. Degree reduction for Bézier curves. *Computer Aided Design*, 20(7):398–405, 1988.

[160] J. Wilczynski. *Projective Differential Geometry of Curves and Ruled Surfaces*. Chelsea, 1961.

[161] A. Worsey, T. Sederberg, and G. Wang. A recursive algortitm for hermite approximation of rational curves, 1994. submitted for publication.

Index

Printed and bound by CPI Group (UK) Ltd, Croydon, CR0 4YY

26/10/2024

01779665-0001